T0185997

Lecture Notes in Mathematics

Volume 2268

Editors-in-Chief
Jean-Michel Morel, CMLA, ENS, Cachan, France
Bernard Teissier, IMJ-PRG, Paris, France

Series Editors
Karin Baur, University of Leeds, Leeds, UK
Michel Brion, UGA, Grenoble, France
Camillo De Lellis, IAS, Princeton, NJ, USA
Alessio Figalli, ETH Zurich, Zurich, Switzerland
Annette Huber, Albert Ludwig University, Freiburg, Germany
Davar Khoshnevisan, The University of Utah, Salt Lake City, UT, USA
Ioannis Kontoyiannis, University of Cambridge, Cambridge, UK
Angela Kunoth, University of Cologne, Cologne, Germany
Ariane Mézard, IMJ-PRG, Paris, France
Mark Podolskij, University of Luxembourg, Esch-sur-Alzette, Luxembourg
Sylvia Serfaty, NYU Courant, New York, NY, USA
Gabriele Vezzosi, UniFI, Florence, Italy
Anna Wienhard, Ruprecht Karl University, Heidelberg, Germany

This series reports on new developments in all areas of mathematics and their applications - quickly, informally and at a high level. Mathematical texts analysing new developments in modelling and numerical simulation are welcome. The type of material considered for publication includes:

1. Research monographs
2. Lectures on a new field or presentations of a new angle in a classical field
3. Summer schools and intensive courses on topics of current research.

Texts which are out of print but still in demand may also be considered if they fall within these categories. The timeliness of a manuscript is sometimes more important than its form, which may be preliminary or tentative.

More information about this series at http://www.springer.com/series/304

Dimitrios Ntalampekos

Potential Theory on Sierpiński Carpets

With Applications to Uniformization

 Springer

Dimitrios Ntalampekos
Department of Mathematics
Stony Brook University
Stony Brook, NY, USA

ISSN 0075-8434   ISSN 1617-9692   (electronic)
Lecture Notes in Mathematics
ISBN 978-3-030-50804-3   ISBN 978-3-030-50805-0   (eBook)
https://doi.org/10.1007/978-3-030-50805-0

Mathematics Subject Classification: Primary: 30L10, 31C45, 46E35; Secondary: 28A75, 30C62, 30C65

© The Editor(s) (if applicable) and The Author(s), under exclusive licence to Springer Nature Switzerland AG 2020
This work is subject to copyright. All rights are solely and exclusively licensed by the Publisher, whether the whole or part of the material is concerned, specifically the rights of translation, reprinting, reuse of illustrations, recitation, broadcasting, reproduction on microfilms or in any other physical way, and transmission or information storage and retrieval, electronic adaptation, computer software, or by similar or dissimilar methodology now known or hereafter developed.
The use of general descriptive names, registered names, trademarks, service marks, etc. in this publication does not imply, even in the absence of a specific statement, that such names are exempt from the relevant protective laws and regulations and therefore free for general use.
The publisher, the authors, and the editors are safe to assume that the advice and information in this book are believed to be true and accurate at the date of publication. Neither the publisher nor the authors or the editors give a warranty, expressed or implied, with respect to the material contained herein or for any errors or omissions that may have been made. The publisher remains neutral with regard to jurisdictional claims in published maps and institutional affiliations.

This Springer imprint is published by the registered company Springer Nature Switzerland AG.
The registered company address is: Gewerbestrasse 11, 6330 Cham, Switzerland

To my parents. . .
   who taught me how to love what I do.

Preface

This work arose during my PhD studies at UCLA. The motivation and the very beginning of this book is a question of my advisor, Mario Bonk: *Can we obtain a uniformization result for Sierpiński carpets using a discrete method?* Such methods, and more specifically the circle packing theorem for triangulations, had already been used by Bonk and Kleiner in their celebrated paper [7], where they provided a sufficient condition for the quasisymmetric uniformization of metric 2-spheres. Another instance was the proof of the Riemann mapping theorem using circle packings by Rodin and Sullivan [35]. Hence, it seemed to be a general principle that discrete uniformization results can be used to obtain uniformization results for a continuum. So I started off working on this problem with an optimistic mindset.

The starting point for mapping a Sierpiński carpet to a square Sierpiński carpet was to mimic Schramm's methods in [36], where he transforms a finite triangulation to a square packing. Soon I realized that there were several hurdles in adapting the technique of Schramm to the carpet setting, which corresponds, in a sense, to an infinite triangulation. There was actually a good reason for that, since behind the difficulties a whole new theory of harmonic functions and quasiconformal maps on carpets was concealed.

Fortunately, at that time I came across the novel work of K. Rajala [33], which had just been posted on arXiv, on the quasiconformal uniformization of metric surfaces. This striking work uses potential theoretic methods in a very general and widely applicable setting to prove a uniformization result. In particular, Rajala reproves the Bonk–Kleiner theorem, as Lytchak and Wenger also did later [26]. Rajala's work provides the "instructions" that one can follow in order to prove some uniformization result and for me it was the perfect roadmap.

The technical difficulties in my proof involved first of all the construction of an appropriate Sobolev space by adapting the existing Sobolev function theory in metric spaces based on *upper gradients* [22], which does not apply though as it stands in the carpet setting. The definition here was based on the notion of *carpet modulus*, developed by Bonk and Merenkov in order to establish a major rigidity result for the standard Sierpiński carpet [9]. Carpet modulus, in turn, is a "descendant" of the *transboundary modulus* of Schramm [37].

After the construction of the Sobolev space, an entirely novel theory of harmonic functions on carpets arose. Chapter 2 of the book is devoted to that theory. The theory is developed up to the great extent that the non-linear potential theoretic methods allow. Several basic properties of harmonic functions are proved towards the uniformization result and beyond. Finally, the developed theory of harmonic functions was used in conjunction with a delicate construction of a conjugate function, which plays the role of a harmonic conjugate, in order to construct a *carpet-quasiconformal* map and prove the uniformization result of Chap. 3.

The entire monograph consists of new results that reflect some of the most recent developments in the field of Analysis on Metric Spaces and in particular in the study of Sierpiński carpets. The material in the greatest possible extent is presented in a self-contained manner. The book is directed towards advanced graduate students and researchers working in the areas of Quasiconformal Geometry, Geometric Group Theory, Complex Dynamics, Geometric Function Theory, Potential Theory, and PDEs. The material in the book can be presented in a semester advanced graduate course as follows. Sections 2.1–2.7 contain the necessary background of harmonic functions for the uniformization result and Chap. 3, with the required preliminaries from Chap. 2 restated, proves the uniformization result. Moreover, this book and the paper of Bonk on the uniformization of carpets [7] can be studied in parallel (for example, in a reading course) and together they provide a complete account on the problem of uniformization of planar Sierpiński carpets.

I would like to thank my advisor, Mario Bonk, for suggesting the study of potential theory on carpets and guiding me throughout this research at UCLA. He has been a true teacher, sharing his expertise in the field and constantly providing deep insight to my questions while being always patient and supportive. Moreover, I thank him for his thorough reading of this work and for his comments and corrections, which substantially improved the presentation. I would also like to thank some people who motivated me to complete this work and especially Pekka Koskela, Kai Rajala, and Jang-Mei Wu. Finally, one anonymous referee, with considerable effort, provided me with valuable feedback, which is much appreciated. The vast majority of the work was completed at UCLA and some minor work involving editing was done at Stony Brook University. I wish to thank the departments of mathematics in both universities for providing an excellent working environment.

The author was partially supported by NSF grant DMS-1506099 during the completion of this research.

Stony Brook, USA   Dimitrios Ntalampekos
May 2020

Contents

Chapter 1
Introduction

One of the main problems in the field of Analysis on Metric Spaces is to find geometric conditions on a given metric space under which the space can be transformed to a "canonical" space with a map that preserves the geometry. In other words, we wish to *uniformize* the metric space by a canonical space. For example, the Riemann mapping theorem gives a conformal map from any simply connected proper subregion of the plane onto the unit disk, which is the canonical space in this case.

In the setting of metric spaces we search instead for other types of maps, such as bi-Lipschitz, quasiconformal or quasisymmetric maps. One method for obtaining such a map is by solving *minimization problems,* such as the problem of minimizing the *Dirichlet energy* $\int_\Omega |\nabla u|^2$ in an open set $\Omega \subset \mathbb{C}$ among Sobolev functions $u \in W^{1,2}(\Omega)$ that have some certain boundary data.

To illustrate the method, we give an informal example. Let $\Omega \subset \mathbb{C}$ be a quadrilateral, i.e., a Jordan region with four marked points on $\partial\Omega$ that define a topological rectangle. Consider two opposite sides $\Theta_1, \Theta_3 \subset \partial\Omega$ of this topological rectangle. We study the following minimization problem: find a function u for which the infimum

$$\inf\left\{ \int_\Omega |\nabla u|^2 : u \in W^{1,2}(\Omega), \, u\big|_{\Theta_1} = 0, \, u\big|_{\Theta_3} - 1 \right\} \tag{1.1}$$

is attained. One can show that a minimizer u with the right boundary values exists and is harmonic on Ω. Let $D(u) = \int_\Omega |\nabla u|^2$ be the Dirichlet energy of u, and $\Theta_2, \Theta_4 \subset \partial\Omega$ be the other opposite sides of the quadrilateral $\partial\Omega$, numbered in a counter-clockwise fashion. Now, we consider the "dual" problem of finding a minimizer v for

$$\inf\left\{ \int_\Omega |\nabla v|^2 : v \in W^{1,2}(\Omega), \, v\big|_{\Theta_2} = 0, \, v\big|_{\Theta_4} = D(u) \right\}.$$

© The Editor(s) (if applicable) and The Author(s), under exclusive licence to Springer Nature Switzerland AG 2020
D. Ntalampekos, *Potential Theory on Sierpiński Carpets*, Lecture Notes in Mathematics 2268, https://doi.org/10.1007/978-3-030-50805-0_1

Again, it turns out that a minimizer v with the right boundary values exists and is harmonic. In fact, v is the harmonic conjugate of u. Then the mapping $f = (u, v)$ is a conformal map from Ω onto the rectangle $(0, 1) \times (0, D(u))$. See [14] for background on classical potential theory and construction of conformal maps with the above procedure.

This example shows that, in the plane, harmonic functions, which are constructed as minimizers of the Dirichlet energy and solve certain boundary value problems, can be very useful in uniformization theory. Namely, there exist more minimization problems whose solution u can be paired with a harmonic conjugate v as above to yield a conformal map (u, v) that transforms a given region to a canonical region. For example, one can prove in this way the Riemann mapping theorem, the uniformization of annuli by round annuli, and the uniformization of planar domains by slit domains; see [14] for the proofs.

A natural question is whether such methods can be used in the abstract metric space setting in order to obtain uniformization results. For this purpose, one would first need a theory of harmonic functions in metric spaces. Harmonic functions have indeed been studied in depth in the abstract metric space setting. Their definition is based on a suitable notion of Sobolev spaces in metric measure spaces. The usual assumptions on the intrinsic geometry of the metric measure space are that it is doubling and supports a Poincaré inequality. Then, harmonic functions are defined as local energy minimizers among Sobolev functions with the same boundary data. We direct the reader to [39] and the references therein for more background. However, to the best of our knowledge, this general theory has not been utilized yet towards a uniformization result. As we see from the planar examples mentioned previously, a crucial ingredient in order to obtain such a result is the existence of a harmonic conjugate in the two-dimensional setting. Constructing harmonic conjugates turns out to be an extremely challenging task in the metric space setting and it is unclear whether the general theory of harmonic functions in metric spaces can provide harmonic conjugates.

Very recently, though, the construction of harmonic conjugates was achieved by K. Rajala [33], who does not use the existing theory of harmonic functions, but builds an appropriate framework in which the construction can take place. In particular, he solves a minimization problem on metric spaces X homeomorphic to $\mathbb{R}^2$, under some geometric assumptions, and this minimization procedure yields a harmonic function u. This function is then paired with a harmonic conjugate v to provide a quasiconformal homeomorphism (u, v) from X to $\mathbb{R}^2$. The construction of a harmonic conjugate, which is one of the most technical parts of Rajala's work, is very powerful. As a corollary, he obtains the Bonk–Kleiner theorem [8], which asserts that a metric sphere that is *Ahlfors 2-regular* and *linearly locally contractible* is quasisymmetrically equivalent to the standard sphere.

Another minimization problem in similar spirit is Plateau's problem; see the book of Courant [14]. This has also recently been extended to the metric space setting [26] and its solution provides canonical quasisymmetric embeddings of a metric space X into $\mathbb{R}^2$, under some geometric assumptions. Thus, [26] provides an alternative proof of the Bonk–Kleiner theorem.

The development of uniformization results for metric spaces homeomorphic to $\mathbb{R}^2$ or to the sphere would provide some insight towards a better understanding of hyperbolic groups whose boundary at infinity is a 2-sphere. A basic problem in geometric group theory is finding relationships between the algebraic properties of a finitely generated group and the geometric properties of its Cayley graph. For each *Gromov hyperbolic* group G one can define an associated metric space called the *boundary at infinity* of G, denoted by $\partial_\infty G$. This metric space is equipped with a family of *visual metrics*. The geometry of $\partial_\infty G$ is very closely related to the asymptotic geometry of the group G. A major conjecture by Cannon [12] is the following: when $\partial_\infty G$ is homeomorphic to the 2-sphere, then G admits a discrete, cocompact, and isometric action on the hyperbolic 3-space $\mathbb{H}^3$. By a theorem of Sullivan [41], this conjecture is equivalent to the following conjecture:

Conjecture 1.1 If G is a Gromov hyperbolic group and $\partial_\infty G$ is homeomorphic to the 2-sphere, then $\partial_\infty G$, equipped with a visual metric, is quasisymmetric to the 2-sphere.

We now continue our discussion on applications of potential theory and minimization problems to uniformization. We provide an example from the discrete world. In [36], using again an energy minimization procedure, Schramm proved the following fact. Let Ω be a quadrilateral, and T a finite triangulation of Ω with vertex set $\{v\}_{v \in I}$. Then there exists a square tiling $\{Z_v\}_{v \in I}$ of a rectangle R such that each vertex v corresponds to a square Z_v, and two squares Z_u, Z_v are in contact whenever the vertices u, v are adjacent in the triangulation. In addition, the vertices corresponding to squares at corners of R are at the corners of the quadrilateral Ω. In other words, triangulations of quadrilaterals can be transformed to square tilings of rectangles. Of course, we are not expecting any metric properties for the correspondence between vertices and squares, since we are not endowing the triangulation with a metric and we are only taking into account the adjacency of vertices.

Hence, it is evident that potential theory is a valuable tool that is also available in metric spaces and can be used to solve uniformization problems. Furthermore, from the aforementioned results we see that harmonic functions and energy minimizers interact with quasiconformal and quasisymmetric maps in metric spaces. We now switch our discussion to Sierpiński carpets and related uniformization problems.

A planar *Sierpiński carpet* $S \subset \mathbb{C}$ is a locally connected continuum with empty interior that arises from a closed Jordan region $\overline{\Omega}$ by removing countably many Jordan regions Q_i, $i \in \mathbb{N}$, from $\overline{\Omega}$ such that the closures $\overline{Q_i}$ are disjoint with each other and with $\partial \Omega$. The local connectedness assumption can be replaced with the assumption that $\mathrm{diam}(Q_i) \to 0$ as $i \to \infty$. The sets $\partial \Omega$ and ∂Q_i for $i \in \mathbb{N}$ are called the *peripheral circles* of the carpet S and the Jordan regions Q_i, $i \in \mathbb{N}$, are called the *peripheral disks*. According to a theorem of Whyburn [44] all such continua are homeomorphic to each other and, in particular, to the *standard Sierpiński carpet*, which is formed by removing the middle square of side length $1/3$ from the unit square $[0, 1]^2$ and then proceeding inductively in each of the remaining eight squares; see Fig. 1.1.

Fig. 1.1 The standard
Sierpińsi carpet

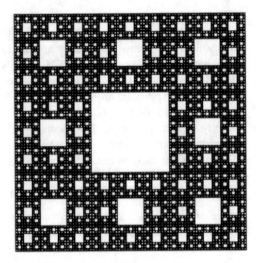

The study of uniformization problems on carpets was initiated by Bonk in [7], where he proved that every Sierpiński carpet in the sphere $\widehat{\mathbb{C}}$ whose peripheral circles are uniform quasicircles and they are also uniformly relatively separated is quasisymmetrically equivalent to a *round* Sierpiński carpet, i.e., a carpet all of whose peripheral circles are geometric circles. The method that he used does not rely on any minimization procedure, but it uses results from complex analysis, and, in particular, Koebe's theorem that allows one to map conformally a finitely connected domain in the plane to a circle domain.

A partial motivation for the development of uniformization results for carpets is another conjecture from geometric group theory, known as the Kapovich–Kleiner conjecture. The conjecture asserts that if a Gromov hyperbolic group G has a boundary at infinity $\partial_\infty G$ that is homeomorphic to a Sierpiński carpet, then G admits a properly discontinuous, cocompact, and isometric action on a convex subset of the hyperbolic 3-space $\mathbb{H}^3$ with non-empty totally geodesic boundary. The Kapovich–Kleiner conjecture [24] is equivalent to the following uniformization problem, similar in spirit to Conjecture 1.1:

Conjecture 1.2 If G is a Gromov hyperbolic group and $\partial_\infty G$ is a Sierpiński carpet, then $\partial_\infty G$ can be quasisymmetrically embedded into the 2-sphere.

The main focus in this work is to prove a uniformization result for planar Sierpiński carpets, under suitable geometric assumptions, by using an energy minimization method. We believe that these methods can be extended to some non-planar carpets and therefore provide some insight to the problem of embedding these carpets into the plane. The canonical spaces in our setting are square carpets, which arise naturally as the extremal spaces of a minimization problem. A *square carpet* here is a planar carpet all of whose peripheral circles are squares, except for the one that separates the rest of the carpet from ∞, which is only required be a rectangle.

Also, the sides of the squares and the rectangle are required to be parallel to the coordinate axes. Our main theorem is the following:

Theorem 1.3 (Theorem 3.1.1) *Let $S \subset \mathbb{C}$ be a Sierpiński carpet of measure zero. Assume that the peripheral disks of S are uniformly Ahlfors 2-regular and uniformly quasiround. Then there exists a homeomorphism from S onto a square carpet that is carpet-quasiconformal.*

The precise definitions of the geometric assumptions and of the notion of quasiconformality that we are employing are given in Chap. 3, Sect. 3.1; see Theorem 3.1.1. Roughly speaking, Ahlfors 2-regularity prevents outward pointing cusps in a uniform way. The quasiroundness assumption says that in large scale the peripheral disks Q_i look like balls, in the sense that for each Q_i there exist two concentric balls, one contained in Q_i and one containing Q_i, with uniformly bounded ratio of radii. For example, if the peripheral disks are John domains with uniform constants, then they satisfy the assumptions; see [40] for the definition of a John domain. The uniformizing map is quasiconformal in the sense that it almost preserves *carpet modulus*, a discrete notion of modulus suitable for Sierpiński carpets. Moreover, under the geometric assumptions of the theorem, this regularity for a uniformizing map onto a square carpet seems to be optimal. Carpet modulus was previously employed by Bonk and Merenkov [9] and is closely related to the notion of *transboundary modulus* of Schramm [37].

If one strengthens the assumptions, then one obtains a quasisymmetry:

Theorem 1.4 (Theorem 3.1.2) *Let $S \subset \mathbb{C}$ be a Sierpiński carpet of measure zero. Assume that the peripheral circles of S are uniformly relatively separated, uniform quasicircles. Then there exists a quasisymmetry from S onto a square carpet.*

These are the same assumptions as the ones used in [7], except for the measure zero assumption, which is essential for our method. The assumption of uniform quasicircles is necessary both in our result and in the uniformization by round carpets result of [7], because this property is preserved under quasisymmetries, and squares and circles share it. The uniform relative separation condition prevents large peripheral circles to be too close to each other. This is essentially the best possible condition one could hope for:

Proposition 1.5 (Proposition 3.1.6) *A round carpet is quasisymmetrically equivalent to a square carpet if and only if the uniform relative separation condition holds.*

The map in Theorem 1.3 is the pair of a certain *carpet-harmonic* function u with its "harmonic conjugate" v. Recall that the carpet S is equal to $\overline{\Omega} \setminus \bigcup_{i \in \mathbb{N}} Q_i$, where Ω is a Jordan region. We wish to view $\partial\Omega$ as a topological rectangle with sides $\Theta_1, \ldots, \Theta_4$ and consider a discrete analog of the minimization problem (1.1). This is the problem that will provide us with the real part u of the uniformizing map. Then, adapting the methods of [33] we construct a "harmonic conjugate" v of u. This is discussed in Chap. 3.

Hence, in order to proceed, we need to introduce a suitable notion of a Sobolev space $\mathcal{W}^{1,2}(S)$ for functions defined on S and a corresponding notion of carpet-harmonic functions. This is the content of Chap. 2.

Before providing a sketch of our definition of Sobolev spaces and carpet-harmonic functions, we recall the definition of Sobolev spaces—also called Newtonian spaces—and harmonic functions on metric spaces, following [38] and [39]. Roughly speaking, a function $u \colon X \to \mathbb{R}$ lies in the Newtonian space $N^{1,p}(X)$ if $u \in L^p(X)$, and there exists a function $g \in L^p(X)$ with the property that

$$|u(x) - u(y)| \le \int_\gamma g \, ds$$

for *almost every* path γ and all points $x, y \in \gamma$. Here, "almost every" means that a family of paths with p-modulus zero has to be excluded; see Sects. 2.3 and 2.4 for a discussion on modulus and non-exceptional paths. The function g is called a *weak upper gradient* of u. Let $I(u) = \inf_g \|g\|_p$ where the infimum is taken over all weak upper gradients of u. A p-harmonic function in an open set $\Omega \subset X$ with boundary data $f \in N^{1,p}(X)$ is a function that minimizes the *energy functional* $I(u)$ over functions $u \in N^{1,p}(X)$ with $u|_{X \setminus \Omega} \equiv f|_{X \setminus \Omega}$. As already remarked, the usual assumptions on the space X for this theory to go through is that it is doubling and supports a Poincaré inequality.

In our setting, we follow a slightly different approach and we do not use measure and integration in the carpet S to study Sobolev functions, but we rather put the focus on studying the "holes" Q_i of the carpet. Hence, we do not make any assumptions on the intrinsic geometry of the carpet S, other than it has Lebesgue measure zero, but we require that the holes Q_i satisfy some uniform geometric conditions; see Sect. 2.2. In particular, we do not assume that the carpet S supports a Poincaré inequality or a doubling measure. What is special about the theory that we develop is that Sobolev functions and harmonic functions will acknowledge in some sense the existence of the ambient space, where the carpet lives.

The precise definitions will be given later in Sects. 2.5 and 2.6 but here we give a rough sketch. A function $u \colon S \to \mathbb{R}$ lies in the Sobolev space $\mathcal{W}^{1,2}(S)$ if it satisfies a certain L^2-integrability condition and it has an *upper gradient* $\{\rho(Q_i)\}_{i \in \mathbb{N}}$, which is a square-summable sequence with the property that

$$|u(x) - u(y)| \le \sum_{i : Q_i \cap \gamma \ne \emptyset} \rho(Q_i)$$

for *almost every* path $\gamma \subset \Omega$ and points $x, y \in \gamma \cap S$. We remark here that the path γ will also travel through the ambient space Ω, and does not stay entirely in the carpet S. Here, "almost every" means that we exclude a family of "pathological" paths of *carpet modulus* equal to zero; see Sect. 2.3 for the definition. This is necessary, because there exist (a lot of) paths γ that are entirely contained in the carpet S without intersecting any peripheral disk $\overline{Q}_i$. For such paths the sum

$\sum_{i:Q_i \cap \gamma \neq \emptyset} \rho(Q_i)$ would be 0, and thus a function u satisfying the upper gradient inequality for *all* paths would be constant.

In order to define a *carpet-harmonic* function, one then minimizes the *energy functional* $\sum_{i \in \mathbb{N}} \rho(Q_i)^2$ over all Sobolev functions that have given boundary data. This energy functional corresponds to the classical Dirichlet energy $\int |\nabla u|^2$ of a classical Sobolev function in the plane.

We will develop this theory for a generalization of Sierpiński carpets called *relative Sierpiński carpets*. The difference with a Sierpiński carpet is that here we will actually allow the set Ω to be an arbitrary (connected) open set in the plane, and not necessarily a Jordan region. So, we start with an open set $\Omega \subset \mathbb{C}$ and we remove the countably many peripheral disks Q_i from Ω as in the definition of a Sierpiński carpet; see Sect. 2.2 for the definition. This should be regarded as a generalization of relative Schottky sets studied in [27], where all peripheral disks Q_i are round disks. This generalization allows us, for example, to set $\Omega = \mathbb{C}$ and obtain an analog of Liouville's theorem, that bounded carpet-harmonic functions are constant.

Under certain assumptions on the geometry of the peripheral disks Q_i (see Sect. 2.2) we obtain the following results (or rather discrete versions of them) for carpet-harmonic functions:

- Solution to the Dirichlet problem; see Sect. 2.6.
- Continuity, maximum principle, uniqueness of the solution to the Dirichlet problem, comparison principle; see Sect. 2.7.
- Caccioppoli's inequality; see Sect. 2.8.
- Harnack's inequality, Liouville's theorem, strong maximum principle; see Sect. 2.9.
- Local equicontinuity and compactness properties of harmonic functions; see Sect. 2.10.

Chapter 2
Harmonic Functions on Sierpiński Carpets

2.1 Introduction

In this chapter we introduce and study notions of Sobolev spaces and harmonic functions on Sierpiński carpets. We briefly describe here some of the applications of carpet-harmonic functions, and then the organization of the current chapter.

In Chap. 3, carpet-harmonic functions are applied towards a uniformization result. In particular, it is proved there that Sierpiński carpets, under the geometric assumptions described in Sect. 2.2, can be uniformized by *square* carpets. This is done by constructing a "harmonic conjugate" of a certain carpet-harmonic function, and modifying the methods used in [33]. The uniformizing map is not quasisymmetric, in general, but it is quasiconformal in a discrete sense. If the assumptions on the peripheral circles are strengthened to *uniformly relatively separated* (see Remark 2.5.4 for the definition), *uniform quasicircles*, then the map is actually a quasisymmetry.

Carpet-harmonic functions are also useful in the study of rigidity problems for quasisymmetric or bi-Lipschitz maps between square Sierpiński carpets. The reason is that the real and imaginary parts of such functions are carpet-harmonic, under some conditions; see Corollary 2.6.5. Such a rigidity problem is studied by Bonk and Merenkov in [9], where it is shown that the only quasisymmetric self-maps of the standard Sierpiński carpet are Euclidean isometries; see [9, Theorem 1.1]. In Theorem 2.7.13 we use the theory of carpet-harmonic functions to show an elementary rigidity result for mappings between square Sierpiński carpets that preserve the sides of the unbounded peripheral disk. This result was already established with different methods in [9, Theorem 1.4]. It would be very interesting to find a proof of the main result of Bonk and Merenkov [9, Theorem 1.1] using carpet-harmonic functions.

The sections of the chapter are organized as follows. In Sect. 2.2 we introduce our notation and our basic assumptions on the geometry of the peripheral disks.

© The Editor(s) (if applicable) and The Author(s), under exclusive licence
to Springer Nature Switzerland AG 2020
D. Ntalampekos, *Potential Theory on Sierpiński Carpets*, Lecture Notes
in Mathematics 2268, https://doi.org/10.1007/978-3-030-50805-0_2

In Sect. 2.3 we discuss notions of *carpet modulus* that will be useful in studying path families in Sierpiński carpets, and, in particular, in defining families of modulus zero which contain "pathological" paths that we wish to exclude from our study. In Sect. 2.4 we prove the existence of paths with certain properties that avoid the "pathological" families of modulus zero.

In Sect. 2.5 we finally introduce Sobolev spaces, starting first with a preliminary notion of a discrete Sobolev function, and then deducing the definition of a Sobolev function. We also study several properties of these functions and give examples.

Section 2.6 discusses the solution to the Dirichlet problem on carpets. Then in Sect. 2.7 we establish several classical properties of harmonic functions, including the continuity, the maximum principle, the uniqueness of the solution to the Dirichlet problem, and the comparison principle. We also prove a discrete analog of the Caccioppoli inequality in Sect. 2.8.

Some more fine properties of carpet-harmonic functions are discussed Sect. 2.9, where we show Harnack's inequality, the analog of Liouville's theorem, and the strong maximum principle. We finish this chapter with Sect. 2.10, where we study equicontinuity and convergence properties of carpet-harmonic functions.

2.2 Basic Assumptions and Notation

We use the notation $\widehat{\mathbb{R}} = \mathbb{R} \cup \{-\infty, +\infty\}$, and $\widehat{\mathbb{C}} = \mathbb{C} \cup \{\infty\}$. A function that attains values in $\widehat{\mathbb{R}}$ is called an *extended function*. We use the standard open ball notation $B(x, r) = \{y \in \mathbb{R}^2 : |x - y| < r\}$ and denote a closed ball by $\overline{B}(x, r)$. If $B = B(x, r)$ then cB denotes the ball $B(x, cr)$. Also, $A(x; r, R)$ denotes the annulus $B(x, R) \setminus \overline{B}(x, r)$, for $0 < r < R$. All the distances will be measured using the Euclidean distance of $\mathbb{C} \simeq \mathbb{R}^2$. The variable x will denote most of the times a point of $\mathbb{R}^2$ and on rare occasion we will use the notation (x, y) for coordinates of a point in $\mathbb{R}^2$, in which case $x, y \in \mathbb{R}$. Each case will be clear from the context.

For $\alpha > 0$ the α-*dimensional Hausdorff measure* $\mathcal{H}^\alpha(E)$ of a set $E \subset \mathbb{C}$ is defined by

$$\mathcal{H}^\alpha(E) = \lim_{\delta \to 0} \mathcal{H}^\alpha_\delta(E) = \sup_{\delta > 0} \mathcal{H}^\alpha_\delta(E),$$

where

$$\mathcal{H}^\alpha_\delta(E) := \inf \left\{ \sum_{j=1}^\infty \operatorname{diam}(U_j)^\alpha : E \subset \bigcup_j U_j, \operatorname{diam}(U_j) < \delta \right\}.$$

The quantity $\mathcal{H}^\alpha_\delta(E)$ is called the α-*dimensional Hausdorff* δ-*content* of E. For $\alpha = 2$, we may normalize $\mathcal{H}^2$ by multiplying it with a constant, so that it agrees with the two-dimensional Lebesgue measure. We will use this normalization throughout the chapter.

Let $\Omega \subset \mathbb{C}$ be a connected open set, and let $\{Q_i\}_{i \in \mathbb{N}}$ be a collection of (open) Jordan regions compactly contained in Ω, with disjoint closures, such that the set $S = \Omega \setminus \bigcup_{i \in \mathbb{N}} Q_i$ has empty interior and is locally connected. The latter will be true if and only if for every ball $B(x, r)$ that is compactly contained in Ω the Jordan regions Q_i with $Q_i \cap B(x, r) \neq \emptyset$ have diameters shrinking to 0. We call the pair (S, Ω) a *relative Sierpiński carpet*. We will often drop Ω from the notation, and just call S a relative Sierpiński carpet. The Jordan regions Q_i are called the *peripheral disks* of S, and the boundaries ∂Q_i are the *peripheral circles*. Note here that $\partial \Omega \cap S = \emptyset$. The definition of a relative Sierpiński carpet is motivated by the fact that if Ω is a Jordan region, then $\overline{S}$ is a Sierpiński carpet in the usual sense, as defined in the Introduction, Chap. 1. See Fig. 2.3 for a Sierpiński carpet, and Fig. 2.4 for a relative Sierpiński carpet, in which Ω has two boundary components.

We will impose some further assumptions on the geometry of the peripheral disks $\{Q_i\}_{i \in \mathbb{N}}$. First, we assume that they are *uniformly quasiround*, i.e., there exists a uniform constant $K_0 \geq 1$ such that for each Q_i there exist concentric balls $B(x, r)$, $B(x, R)$ with the property that

$$B(x, r) \subset Q_i \subset B(x, R) \tag{2.1}$$

and $R/r \leq K_0$. The terminology quasiround appears in earlier works of Tyson [42] and Bonk [7].

Second, we assume that the peripheral disks are *uniformly Ahlfors 2-regular*, i.e., there exists a uniform constant $K_1 > 0$ such that for every Q_i and for every ball $B(x, r)$ centered at some $x \in Q_i$ with $r < \mathrm{diam}(Q_i)$ we have

$$\mathcal{H}^2(B(x, r) \cap Q_i) \geq K_1 r^2. \tag{2.2}$$

A Jordan curve $J \subset \mathbb{R}^2$ is a *K-quasicircle* for some $K > 0$, if for any two points $x, y \in J$ there exists an arc $\gamma \subset J$ with endpoints x, y such that $|x - y| \leq K \, \mathrm{diam}(\gamma)$. Note that if the peripheral circles ∂Q_i are uniform quasicircles (i.e., K-quasicircles with the same constant K), then they are both uniformly quasiround and uniformly Ahlfors 2-regular sets. A proof of the first claim can be found in [7, Proposition 4.3] and the second claim is proved in [37, Corollary 2.3], where the notion of an Ahlfors 2-regular set appeared for the first time in the study of conformal maps; Schramm used the terminology *fat* instead of Ahlfors 2-regular. Another example of Jordan regions being quasiround and Ahlfors 2-regular are *John domains*; see [40] for the definition and properties of John domains. We remark that the boundary of such a domain has strictly weaker properties than those of a quasicircle.

The relative Sierpiński carpets that are considering in this chapter will be assumed to satisfy certain of the above geometric assumptions:

Definition 2.2.1 Let S be a relative Sierpiński carpet. We say that S satisfies the *standard assumptions* if:

(a) the peripheral disks of S are uniformly quasiround,
(b) the peripheral disks of S are uniformly Ahlfors 2-regular, and
(c) $\mathcal{H}^2(S) = 0$.

We say that a constant $c > 0$ *depends on the data of the carpet S*, if it depends only on the quasiroundness and Ahlfors regularity constants K_0 and K_1, respectively.

The notation $V \subset\subset \Omega$ means that $\overline{V}$ is compact and is contained in Ω. Alternatively, we say that V *is compactly contained in* Ω. For a set $E \subset \mathbb{R}^2$ and $\delta > 0$ we denote by $N_\delta(E)$ the open δ-neighborhood of E

$$\{x \in \mathbb{R}^2 : \operatorname{dist}(x, E) < \delta\}.$$

A *continuum* $E \subset \mathbb{R}^2$ is a compact and connected set. A continuum E is *non-trivial* if it contains at least two points. Making slight abuse of notation and for visual purposes, we use $\overline{Q_i}$ to denote the closure of Q_i, instead of using $\overline{Q_i}$.

A *path* or *curve* γ is a continuous function $\gamma: I \to \mathbb{R}^2$, where $I \subset \mathbb{R}$ is a bounded interval, such that γ has a continuous extension $\overline{\gamma} : \overline{I} \to \mathbb{R}^2$, i.e., γ has endpoints. A *closed* path γ is a path with $I = [0, 1]$ and an *open* path γ is a path with $I = (0, 1)$. We will also use the notation $\gamma \subset \mathbb{R}^2$ for the image of the path as a set. A *subpath* or *subcurve* of a path $\gamma: I \to \mathbb{R}^2$ is the restriction of γ to a subinterval of I. A *Jordan curve* is a homeomorphic image of the unit circle S^1, and a *Jordan arc* is homeomorphic to $[0, 1]$.

We denote by S° the points of the relative Sierpiński carpet S that do not lie on any peripheral circle ∂Q_i. For an open set $V \subset \Omega$ define $\partial_* V = S \cap \partial V$; see Fig. 2.3. For a set V that intersects the relative Sierpiński carpet S we define the index set $I_V = \{i \in \mathbb{N} : Q_i \cap V \neq \emptyset\}$.

In the proofs we will denote constants by $C, C', C'', \ldots$, where the same symbol can denote a different constant if there is no ambiguity.

2.3 Notions of Carpet Modulus

The carpet modulus is a generalization of the transboundary modulus introduced by Schramm in [37]. We will define two notions of carpet modulus: the weak and strong. The weak carpet modulus was first introduced in [9, Sect. 2], where several of its properties were established.

Let (S, Ω) be a relative Sierpiński carpet with the standard assumptions (see Definition 2.2.1), and let Γ be a family of paths in Ω.

Let us recall first the definition of *conformal modulus* or *2-modulus* of a path family Γ in Ω. A non-negative Borel extended function λ on Ω is *admissible* for the conformal modulus $\mathrm{mod}_2(\Gamma)$ if

$$\int_\gamma \lambda \, ds \geq 1$$

for all locally rectifiable paths $\gamma \in \Gamma$. If a path γ is not locally rectifiable, we define $\int_\gamma \lambda \, ds = \infty$, even when $\lambda \equiv 0$. Hence, we may require that the above inequality holds for all $\gamma \in \Gamma$. Then $\mathrm{mod}_2(\Gamma) = \inf \int \lambda^2 \, d\mathcal{H}^2$ where the infimum is taken over all admissible functions.

A sequence of non-negative numbers $\{\rho(Q_i)\}_{i\in\mathbb{N}}$ is admissible for the *weak (carpet) modulus* $\mathrm{mod}_w(\Gamma)$ if there exists a path family $\Gamma_0 \subset \Gamma$ with $\mathrm{mod}_2(\Gamma_0) = 0$ such that

$$\sum_{i:Q_i\cap\gamma\neq\emptyset} \rho(Q_i) \geq 1 \tag{2.3}$$

for all $\gamma \in \Gamma \setminus \Gamma_0$. Note that in the sum each peripheral disk is counted at most once, and we only include the peripheral disks whose interior is intersected by γ, and not just the boundary. Then we define $\mathrm{mod}_w(\Gamma) = \inf \sum_{i\in\mathbb{N}} \rho(Q_i)^2$ where the infimum is taken over all admissible weights ρ.

Similarly we define the notion of *strong (carpet) modulus*. A sequence of non-negative numbers $\{\rho(Q_i)\}_{i\in\mathbb{N}}$ is admissible for the *strong carpet modulus* $\mathrm{mod}_s(\Gamma)$ if

$$\sum_{i:Q_i\cap\gamma\neq\emptyset} \rho(Q_i) \geq 1 \tag{2.4}$$

for all $\gamma \in \Gamma$ that satisfy $\mathcal{H}^1(\gamma \cap S) = 0$. Note that the path γ could be non-rectifiable inside some Q_i. Then $\mathrm{mod}_s(\Gamma) := \inf \sum_{i\in\mathbb{N}} \rho(Q_i)^2$ where the infimum is taken over all admissible weights ρ.

For properties of the conformal modulus see [25, Sect. 4.2, p. 133]. It can be shown as in the conformal case that both notions of carpet modulus satisfy monotonicity and countable subadditivity, i.e., if $\Gamma_1 \subset \Gamma_2$ then $\mathrm{mod}(\Gamma_1) \leq \mathrm{mod}(\Gamma_2)$ and

$$\mathrm{mod}\left(\bigcup_{i\in\mathbb{N}}\Gamma_i\right) \leq \sum_{i\in\mathbb{N}} \mathrm{mod}(\Gamma_i),$$

where mod is either mod_w or mod_s.

The following lemma provides some insight for the relation between the two notions of carpet modulus.

Lemma 2.3.1 *For any path family Γ in Ω we have*

$$\mathrm{mod}_w(\Gamma) \leq \mathrm{mod}_s(\Gamma).$$

Proof Let $\{\rho(Q_i)\}_{i\in\mathbb{N}}$ be admissible for $\mathrm{mod}_s(\Gamma)$, so $\sum_{i:Q_i\cap\gamma\neq\emptyset}\rho(Q_i) \geq 1$ for all $\gamma \in \Gamma$ with $\mathcal{H}^1(\gamma \cap S) = 0$. Define $\Gamma_0 = \{\gamma \in \Gamma : \mathcal{H}^1(\gamma \cap S) > 0\}$. Then the function $\lambda = \infty \cdot \chi_S$ is admissible for $\mathrm{mod}_2(\Gamma_0)$. Since $\mathcal{H}^2(S) = 0$, it follows that $\mathrm{mod}_2(\Gamma_0) = 0$. Hence, $\sum_{i:Q_i\cap\gamma\neq\emptyset}\rho(Q_i) \geq 1$ for all $\gamma \in \Gamma \setminus \Gamma_0$, which shows the admissibility of $\{\rho(Q_i)\}_{i\in\mathbb{N}}$ for the weak carpet modulus $\mathrm{mod}_w(\Gamma)$. $\qquad\square$

A version of the next lemma can be found in [9, Lemma 2.2] and [6].

Lemma 2.3.2 *Let $\kappa \geq 1$ and I be a countable index set. Suppose that $\{B_i\}_{i\in I}$ is a collection of balls in $\mathbb{R}^2$, and a_i, $i \in I$, are non-negative real numbers. Then there exists a constant $C > 0$ depending only on κ such that*

$$\left\|\sum_{i\in I} a_i \chi_{\kappa B_i}\right\|_2 \leq C\left\|\sum_{i\in I} a_i \chi_{B_i}\right\|_2.$$

Here $\|\cdot\|_2$ denotes the L^2-norm with respect to planar Lebesgue measure. We will also need the next lemma.

Lemma 2.3.3 *For a path family Γ in Ω we have the equivalence*

$$\mathrm{mod}_w(\Gamma) = 0 \quad \text{if and only if} \quad \mathrm{mod}_2(\Gamma) = 0.$$

Before starting the proof, we require the following consequence of the Ahlfors regularity assumption.

Lemma 2.3.4 *The following statements are true.*

(a) *Let E be a compact subset of $\mathbb{R}^2$. Then for each $\varepsilon > 0$ there exist at most finitely many peripheral disks Q_i intersecting E with diameter larger than ε.*
(b) *The spherical diameters of the peripheral disks Q_i of S converge to 0.*

Proof Note first that by an area argument no ball $B(0, R)$ can *contain* infinitely many peripheral disks Q_i with $\mathrm{diam}(Q_i) > \varepsilon > 0$. Indeed, the Ahlfors 2-regularity condition implies that

$$\mathcal{H}^2(Q_i) \geq K_1 \mathrm{diam}(Q_i)^2.$$

Since the peripheral disks Q_i are disjoint we have

$$\mathcal{H}^2(B(0, R)) \geq \sum_{i:Q_i\subset B(0,R)} \mathcal{H}^2(Q_i) \geq K_1 \sum_{i:Q_i\subset B(0,R)} \mathrm{diam}(Q_i)^2.$$

Hence, we see that only finitely many of them can satisfy $\mathrm{diam}(Q_i) > \varepsilon$.

If there exist infinitely many Q_i intersecting E with $\mathrm{diam}(Q_i) > \varepsilon$ then by the above they cannot be contained in any ball $B(0, R)$. In particular there exists a ball $B(0, R) \supset E$ such that there are infinitely many Q_i intersecting both $\partial B(0, R)$ and $\partial B(0, 2R)$. The Ahlfors regularity assumption implies now that for all such Q_i we have

$$\mathcal{H}^2(Q_i \cap (B(0, 2R) \setminus B(0, R))) \geq CR^2 \geq C\varepsilon^2$$

for a constant $C > 0$ depending only on K_1; see Remark 2.3.5 below. Hence, an area argument as before yields the conclusion. This completes the proof of (a).

For (b), suppose that there exists an infinite sequence of peripheral disks with spherical diameters bounded below away from 0. Then there exists a compact set $E \subset \mathbb{R}^2$ intersecting infinitely many of these peripheral disks. In a neighborhood of E the spherical metric is comparable to the Euclidean metric, hence there are infinitely many peripheral disks intersecting E with Euclidean diameters bounded below. This contradicts (a). $\qquad\square$

Remark 2.3.5 In the preceding proof we used the fact that if $\overline{Q}_i$ intersects two circles $\partial B(x, r)$ and $\partial B(x, R)$ with $0 < r < R$, then

$$\mathcal{H}^2(Q_i \cap (B(x, R) \setminus B(x, r))) \geq C(R - r)^2$$

for a constant $C > 0$ depending only on K_1. To see that, by the connectedness of Q_i there exists a point $y \in Q_i \cap \partial B(x, (r + R)/2)$. Then $B(y, (R - r)/2) \subset B(x, R) \setminus B(x, r)$, so

$$\mathcal{H}^2(Q_i \cap (B(x, R) \setminus B(x, r))) \geq \mathcal{H}^2(Q_i \cap B(y, (R - r)/2)) \geq K_1 \frac{(R - r)^2}{4},$$

by the Ahlfors regularity condition (2.2).

Corollary 2.3.6 *Let E be a compact subset of $\mathbb{R}^2$. Then*

$$\sum_{i \in I_E} \mathrm{diam}(Q_i)^2 < \infty.$$

Recall that $I_E = \{i \in \mathbb{N} : Q_i \cap E \neq \emptyset\}$.

Proof Let $B(0, R)$ be a large ball that contains E. Then there are finitely many peripheral disks intersecting $B(0, R)$ and having diameter greater than $R/2$, by Lemma 2.3.4. Hence, it suffices to show

$$\sum_{i : Q_i \subset B(0, 2R)} \mathrm{diam}(Q_i)^2 < \infty.$$

Using the Ahlfors 2-regularity, one can see that this sum is bounded above by a multiple of $\mathcal{H}^2(B(0, 2R))$, as in the proof of Lemma 2.3.4.   □

Proof of Lemma 2.3.3 One direction is trivial, namely if $\mathrm{mod}_2(\Gamma) = 0$ then $\mathrm{mod}_w(\Gamma) = 0$, since the weight $\rho(Q_i) \equiv 0$ is admissible.

For the converse, note first that Γ cannot contain constant paths if $\mathrm{mod}_w(\Gamma) = 0$. Indeed, assume that Γ contains a constant path $\gamma = x_0 \in \Omega$, and $\{\rho(Q_i)\}_{i \in \mathbb{N}}$ is an admissible weight for $\mathrm{mod}_w(\Gamma)$. Then there exists an exceptional family $\Gamma_0 \subset \Gamma$ with $\mathrm{mod}_2(\Gamma_0) = 0$ such that

$$\sum_{i : Q_i \cap \gamma \neq \emptyset} \rho(Q_i) \geq 1$$

for all $\gamma \in \Gamma \setminus \Gamma_0$. The constant path $\gamma = x_0$ cannot lie in Γ_0, otherwise we would have $\mathrm{mod}_2(\Gamma_0) = \infty$ because no function would be admissible for $\mathrm{mod}_2(\Gamma_0)$. Hence, we must have

$$\sum_{i : x_0 \in Q_i} \rho(Q_i) \geq 1.$$

If $x_0 \in S$, then this cannot happen since the sum is empty, so the only possibility is that $x_0 \in Q_{i_0}$ for some $i \in \mathbb{N}$. In this case we have $\rho(Q_{i_0}) \geq 1$. Hence, $\sum_{i \in \mathbb{N}} \rho(Q_i)^2 \geq 1$, which implies that $\mathrm{mod}_w(\Gamma) \geq 1$, a contradiction.

We now proceed to showing the implication. By the subadditivity of 2-modulus, it suffices to show that the family Γ_δ of paths in Γ that have diameter bounded below by $\delta > 0$ has conformal modulus zero. Indeed, this will exhaust all paths of Γ, since Γ contains no constant paths. For simplicity we write Γ in the place of Γ_δ for the remainder of the proof. Note that we have $\mathrm{mod}_w(\Gamma) = 0$, using the monotonicity of modulus.

For $\varepsilon > 0$ let $\{\rho(Q_i)\}_{i \in \mathbb{N}}$ be a weight such that $\sum_{i \in \mathbb{N}} \rho(Q_i)^2 < \varepsilon$ and

$$\sum_{i : Q_i \cap \gamma \neq \emptyset} \rho(Q_i) \geq 1$$

for $\gamma \in \Gamma \setminus \Gamma_0$, where Γ_0 is a path family with $\mathrm{mod}_2(\Gamma_0) = 0$. By using $\varepsilon = 1/2^n$, $n \in \mathbb{N}$, summing the corresponding weights ρ, and taking the union of the exceptional families Γ_0, we obtain a weight $\{\rho(Q_i)\}_{i \in \mathbb{N}}$ and an exceptional family Γ_0 such that $\sum_{i \in \mathbb{N}} \rho(Q_i)^2 < \infty$ and

$$\sum_{i : Q_i \cap \gamma \neq \emptyset} \rho(Q_i) = \infty \tag{2.5}$$

for all $\gamma \in \Gamma \setminus \Gamma_0$, where $\mathrm{mod}_2(\Gamma_0) = 0$.

We construct an admissible function $\lambda \colon \mathbb{C} \to [0, \infty]$ for $\mathrm{mod}_2(\Gamma)$ as follows. Since the peripheral disks Q_i are uniformly quasiround, there exist balls $B(x_i, r_i) \subset Q_i \subset B(x_i, R_i)$ with $R_i/r_i \leq K_0$. We define

$$\lambda = \sum_{i \in \mathbb{N}} \frac{\rho(Q_i)}{R_i} \chi_{B(x_i, 2R_i)}.$$

Note that if γ intersects some Q_i with $4R_i < \delta$, then γ must exit $B(x_i, 2R_i)$, so $\int_\gamma \chi_{B(x_i, 2R_i)} \, ds \geq R_i$. If γ is a bounded path, i.e., it is contained in a ball $B(0, R)$, then there are only finitely many peripheral disks Q_i intersecting γ and satisfying $4R_i \geq \delta$. This follows from Lemma 2.3.4 and the fact that $\mathrm{diam}(Q_i) \geq r_i \geq R_i/K_0$ from the quasiroundness assumption. Thus, we have

$$\sum_{\substack{i : Q_i \cap \gamma \neq \emptyset \\ 4R_i \geq \delta}} \rho(Q_i) < \infty,$$

since it is a finite sum. This implies that

$$\int_\gamma \lambda \, ds \geq \sum_{\substack{i : Q_i \cap \gamma \neq \emptyset \\ 4R_i < \delta}} \frac{\rho(Q_i)}{R_i} \int_\gamma \chi_{B(x_i, 2R_i)} \, ds \geq \sum_{\substack{i : Q_i \cap \gamma \neq \emptyset \\ 4R_i < \delta}} \rho(Q_i) = \infty$$

by (2.5), whenever $\gamma \in \Gamma \setminus \Gamma_0$. Now, if $\gamma \in \Gamma \setminus \Gamma_0$ is an unbounded path, then γ always exits $B(x_i, 2R_i)$ whenever $Q_i \cap \gamma \neq \emptyset$, so in this case we also have

$$\int_\gamma \lambda \, ds = \infty.$$

Using Lemma 2.3.2 we obtain

$$\|\lambda\|_2 \leq C \left\| \sum_{i \in \mathbb{N}} \frac{\rho(Q_i)}{R_i} \chi_{B(x_i, R_i/K_0)} \right\|_2 \leq C \left\| \sum_{i \in \mathbb{N}} \frac{\rho(Q_i)}{R_i} \chi_{B(x_i, r_i)} \right\|_2$$

$$\leq C' \left(\sum_{i \in \mathbb{N}} \rho(Q_i)^2 \right)^{1/2} < \infty.$$

since the balls $B(x_i, r_i)$ are disjoint. This implies that $\mathrm{mod}_2(\Gamma \setminus \Gamma_0) = 0$. Thus,

$$\mathrm{mod}_2(\Gamma) \leq \mathrm{mod}_2(\Gamma \setminus \Gamma_0) + \mathrm{mod}_2(\Gamma_0) = 0$$

by the subadditivity of modulus. $\qquad\square$

Remark 2.3.7 Observe that families of paths passing through a single point $p \in Q_i$ would have conformal modulus and thus weak carpet modulus equal to zero, but this is not the case when we use the strong modulus. The strong modulus would not distinguish in this case the family of paths passing through Q_i from the family passing just through a point $p \in Q_i$. In what follows we will study in parallel the two notions, pointing out the differences whenever they occur.

Finally, we recall Fuglede's lemma in this setting:

Lemma 2.3.8 *Let $\{\rho(Q_i)\}_{i \in \mathbb{N}}$ and $\{\rho_n(Q_i)\}_{i \in \mathbb{N}}$ for $n \in \mathbb{N}$ be non-negative weights in $\ell^2(\mathbb{N})$ such that $\rho_n \to \rho$ in $\ell^2(\mathbb{N})$, i.e.,*

$$\sum_{i \in \mathbb{N}} |\rho_n(Q_i) - \rho(Q_i)|^2 \to 0$$

as $n \to \infty$. Then there exists a subsequence $\{\rho_{k_n}(Q_i)\}_{i \in \mathbb{N}}$, $n \in \mathbb{N}$, and an exceptional family Γ_0 with $\mathrm{mod}_s(\Gamma_0) = 0$ such that for all paths $\gamma \subset \Omega$ with $\gamma \notin \Gamma_0$ we have

$$\sum_{i: Q_i \cap \gamma \neq \emptyset} |\rho_{k_n}(Q_i) - \rho(Q_i)| \to 0$$

as $n \to \infty$.

The proof is a simple adaptation of the conformal modulus proof but we include it here for the sake of completeness. The argument is essentially contained in the proof of [9, Proposition 2.4, pp. 604–605].

Proof Without loss of generality, we may assume that $\rho_n \geq 0$ and $\rho_n \to 0$ in $\ell^2(\mathbb{N})$. We consider a subsequence ρ_{k_n} such that

$$\sum_{i \in \mathbb{N}} \rho_{k_n}(Q_i)^2 < \frac{1}{2^n}$$

for all $n \in \mathbb{N}$. By the subadditivity of strong modulus, it suffices to show that for each $\delta > 0$ the path family

$$\Gamma_0 := \{\gamma \subset \Omega : \limsup_{n \to \infty} \sum_{i: Q_i \cap \gamma \neq \emptyset} \rho_{k_n}(Q_i) > \delta\}$$

has strong modulus zero.

Let

$$\lambda = \sum_{n=1}^{\infty} \rho_{k_n}$$

and note that

$$\sum_{i: Q_i \cap \gamma \neq \emptyset} \lambda(Q_i) = \sum_{n=1}^{\infty} \sum_{i: Q_i \cap \gamma \neq \emptyset} \rho_{k_n}(Q_i) = \infty$$

for all $\gamma \in \Gamma_0$. On the other hand,

$$\left(\sum_{i \in \mathbb{N}} \lambda(Q_i)^2\right)^{1/2} = \|\lambda\|_{\ell^2(\{Q_i : i \in \mathbb{N}\})} = \left\|\sum_{n=1}^{\infty} \rho_{k_n}\right\|_{\ell^2(\{Q_i : i \in \mathbb{N}\})}$$

$$\leq \sum_{n=1}^{\infty} \|\rho_{k_n}\|_{\ell^2(\{Q_i : i \in \mathbb{N}\})} \leq \sum_{n=1}^{\infty} \frac{1}{2^n} < \infty.$$

Since $\varepsilon \cdot \lambda$ is admissible for $\mathrm{mod}_s(\Gamma_0)$ for all $\varepsilon > 0$, it follows that $\mathrm{mod}_s(\Gamma_0) = 0$. The proof is complete. $\qquad\square$

2.4 Existence of Paths

In this section we will show the existence of paths that avoid given families of (weak, strong, conformal) modulus equal to 0. These paths will therefore be "good" paths for which we can apply, e.g., Fuglede's lemma. We will use these good paths later to prove qualitative estimates, such as continuity of carpet-harmonic functions.

First we recall some facts. The *co-area inequalities* in the next proposition are modified versions of [17, Theorem 3.2.12] and [1, Proposition 3.1.5].

Proposition 2.4.1 *Let* $T : \mathbb{R}^2 \to \mathbb{R}$ *be an L-Lipschitz function and* g *be a non-negative measurable function on* $\mathbb{R}^2$. *Then the function* $x \mapsto \int_{T^{-1}(x)} g(y) \, d\mathcal{H}^1(y)$ *is measurable, and there is a constant* $C > 0$ *depending only on* L *such that:*

$$\int_{\mathbb{R}} \left(\int_{T^{-1}(x)} g(z) \, d\mathcal{H}^1(z)\right) dx \leq C \int_{\mathbb{R}^2} g(z) \, d\mathcal{H}^2(z), \qquad \text{(Co-area inequality)}$$

and

$$\int_{\mathbb{R}} \sum_{z \in T^{-1}(x)} g(z) \, dx \leq C \int_{\mathbb{R}^2} g(z) \, d\mathcal{H}^1(z). \qquad \text{(Co-area inequality)}$$

We say that a path α *joins* or *connects* two sets E, F if $\overline{\alpha}$ intersects both E and F. The following proposition asserts that perturbing a curve yields several nearby curves; see [10, Theorem 3].

Proposition 2.4.2 *Let* $\alpha \subset \mathbb{R}^2$ *be a closed path that joins two non-trivial, disjoint continua* $E, F \subset \mathbb{R}^2$. *Consider the distance function* $\psi(x) = \mathrm{dist}(x, \alpha)$. *Then there exists* $\delta > 0$ *such that for a.e.* $s \in (0, \delta)$ *there exists a simple path* $\alpha_s \subset \psi^{-1}(s)$ *joining* E *and* F.

Using that, we prove the following lemma.

Lemma 2.4.3 *Let $\alpha \subset \Omega$ be a closed path joining two non-trivial, disjoint continua E, $F \subset\subset \Omega$, and let Γ be a given family of (weak, strong, conformal) modulus zero. Then, there exists $\delta > 0$ such that for a.e. $s \in (0, \delta)$ there exists a simple path $\alpha_s \subset \psi^{-1}(s)$ that lies in Ω, joins the continua E and F, and lies outside the family Γ. Furthermore, if $A \subset \Omega$ is a given set with $\mathcal{H}^1(A) = 0$, then for a.e. $s \in (0, \delta)$ the path α_s does not intersect A.*

Proof Note that if Γ has strong or weak modulus zero, then it actually has conformal modulus zero, by Lemmas 2.3.1 and 2.3.3. Hence, it suffices to assume that $\mathrm{mod}_2(\Gamma) = 0$.

For $\varepsilon > 0$ there exists an admissible function λ such that $\int_\gamma \lambda\, ds \geq 1$ for all $\gamma \in \Gamma$, and $\|\lambda\|_2 < \varepsilon$. Consider a small $\delta > 0$ such that $N_\delta(\alpha) \subset\subset \Omega$ and the conclusion of Proposition 2.4.2 is true. Let J be the set of $s \in (0, \delta)$ such that $\alpha_s \in \Gamma$, and J' be the set of $s \in (0, \delta)$ such that $\int_{\psi^{-1}(s)} \lambda\, d\mathcal{H}^1 \geq 1$. It is clear that $J \subset J'$, and J' is measurable by Proposition 2.4.1, since the function ψ is 1-Lipschitz. Thus, applying the first co-area inequality in Proposition 2.4.1 and the Cauchy–Schwarz inequality, we have

$$\mathcal{H}^1(J) \leq \mathcal{H}^1(J') = \int_{J'} d\mathcal{H}^1(s) \leq \int_{J'} \left(\int_{\psi^{-1}(s)} \lambda\, d\mathcal{H}^1 \right) d\mathcal{H}^1(s)$$

$$\leq C \int_{N_\delta(\alpha)} \lambda\, d\mathcal{H}^2 \leq C\mathcal{H}^2(N_\delta(\alpha))^{1/2}\|\lambda\|_2 < C'\varepsilon.$$

Letting $\varepsilon \to 0$, we obtain $\mathcal{H}^1(J) = 0$, and this completes the proof.

Finally, we show the latter claim. Here we will use the second co-area inequality in Proposition 2.4.1. For $g(z) = \chi_A \chi_{N_\delta(\alpha)}$ we have

$$\int_0^\delta \#\{\psi^{-1}(s) \cap A\}\, ds \leq C \int_{N_\delta(\alpha)} \chi_A\, d\mathcal{H}^1 = 0.$$

Here, $\#$ is the counting measure. Hence, $\#\{\psi^{-1}(s) \cap A\} = 0$ for a.e. $s \in (0, \delta)$, and the conclusion for $\alpha_s \subset \psi^{-1}(s)$ follows immediately. $\square$

We also need a version of the above lemma for paths intersecting the boundary of Ω:

Lemma 2.4.4 *Let $\alpha \subset \Omega$ be an open path with $\overline{\alpha} \cap \partial\Omega \neq \emptyset$, and let Γ be a given family of (weak, strong, conformal) modulus zero. Assume that $x \in \overline{\alpha} \cap \partial\Omega$ lies in a non-trivial component of $\partial\Omega$. Then, for every $\varepsilon > 0$ there exists a $\delta > 0$ such that for a.e. $s \in (0, \delta)$ there exists an open path $\alpha_s \subset \psi^{-1}(s)$ that lies in Ω, lands at a point $x_s \in B(x, \varepsilon) \cap \partial\Omega$, and avoids the path family Γ. Furthermore, if $A \subset \Omega$ is a given set with $\mathcal{H}^1(A) = 0$, then for a.e. $s \in (0, \delta)$ the path α_s does not intersect A.*

Proof We only sketch the part of the proof related to the landing point of α_s, since the rest is the same as the proof of Lemma 2.4.3.

Note that for small $\varepsilon > 0$ there exists a connected subset E of $\partial\Omega$ that connects x to $\partial B(x, \varepsilon)$. Hence, if we apply Lemma 2.4.2 to the path $\overline{\alpha}$ we can obtain paths in $\psi^{-1}(s) \subset \mathbb{R}^2$ that land at E; here $F \subset \Omega$ can be any continuum that intersects $\overline{\alpha}$. However, these paths do not lie necessarily in Ω, so in this case we have to truncate them at the "first time" that they meet $\partial\Omega$, assuming that they are parametrized so that they start at F and end at E. If $\delta > 0$ is chosen sufficiently small, then for a.e. $s \in (0, \delta)$ these paths will land at a point $x_s \in B(x, \varepsilon) \cap \partial\Omega$. □

Next, we switch to a special type of curves α_s, namely circular arcs. A sequence of weights $\{h(Q_i)\}_{i \in \mathbb{N}}$ is *locally square-summable* if for each $x \in S$ there exists a ball $B(x, r) \subset \Omega$ such that

$$\sum_{i \in I_B} h(Q_i)^2 < \infty.$$

Remark 2.4.5 Let $\{h(Q_i)\}_{i \in \mathbb{N}}$ be a sequence of non-negative weights such that $\sum_{i \in \mathbb{N}} h(Q_i) < \infty$, and let $x \in S^\circ \cup \partial\Omega$, i.e., x does not lie on the boundary of any peripheral disk. Then

$$\sum_{i \in I_{B(x,r)}} h(Q_i) \to 0$$

as $r \to 0$. This is because for any given peripheral disk Q_i, the intersection of Q_i with $B(x, r)$ is empty for all sufficiently small $r > 0$.

Remark 2.4.6 Let Γ be a path family in $\mathbb{R}^2$ with (weak, strong, conformal) modulus zero. Then the family Γ_0 of paths in $\mathbb{R}^2$ that contain a subpath lying in Γ also has (weak, strong, conformal) modulus zero.

Lemma 2.4.7 *Let $\{h(Q_i)\}_{i \in \mathbb{N}}$ be a locally square-summable sequence. Consider a set $A \subset \Omega$ with $\mathcal{H}^1(A) = 0$, and a path family Γ in Ω that has (weak, strong) modulus equal to zero.*

(a) *If $x \in S^\circ \cup \partial\Omega$ then for each $\varepsilon > 0$ we can find an arbitrarily small $r > 0$ such that the circular path $\gamma_r(t) = x + re^{it}$, as well as all of its subpaths, does not lie in Γ, it does not intersect A, and*

$$\sum_{i:Q_i \cap \gamma_r \neq \emptyset} h(Q_i) < \varepsilon.$$

If $x \in \partial Q_{i_0}$ for some $i_0 \in \mathbb{N}$, then the same conclusion is true, if we exclude the peripheral disk Q_{i_0} from the above sum.

(b) *If $x, y \in \partial Q_{i_0}$, then for each $\varepsilon > 0$ we can find an arbitrarily small $r > 0$ and a path $\gamma_0 \subset Q_{i_0}$ that joins the circular paths $\gamma_r^x(t) = x + re^{it}$, $\gamma_r^y(t) = y + re^{it}$*

with the following property: any simple path γ contained in the concatenation of the paths γ_0, γ_r^x, γ_r^y does not lie in Γ, γ does not intersect A, and

$$\sum_{\substack{i:Q_i\cap\gamma\neq\emptyset \\ i\neq i_0}} h(Q_i) \leq \sum_{\substack{i:Q_i\cap(\gamma_r^x\cup\gamma_r^y)\neq\emptyset \\ i\neq i_0}} h(Q_i) < \varepsilon.$$

Proof We may assume that $\mathrm{mod}_2(\Gamma) = 0$, by Lemmas 2.3.1 and 2.3.3.

(a) Note that the circular path γ_r centered at x lies in the set $\psi^{-1}(r)$, where $\psi(z) = |z - x|$ is a 1-Lipschitz function. As in the proof of Lemma 2.4.3 one can show that there exists $\delta > 0$ such that for a.e. $r \in (0, \delta)$ the path γ_r avoids Γ and the set A. Remark 2.4.6 implies that all subpaths of γ_r also avoid Γ. Assume that $x \in S^\circ \cup \partial\Omega$ and fix $\varepsilon > 0$. In order to show the statement, it suffices to show that for arbitrarily small $\delta > 0$, there exists a set $J \subset (\delta/2, \delta)$ of positive measure such that

$$\sum_{i:Q_i\cap\gamma_r\neq\emptyset} h(Q_i) < \varepsilon$$

for all $r \in J$. Assume that this fails, so there exists a small $\delta > 0$ such that the reverse inequality holds for a.e. $r \in (\delta/2, \delta)$. Noting that the function $r \mapsto \chi_{Q_i\cap\gamma_r}$ is measurable and integrating over $r \in (\delta/2, \delta)$, we obtain

$$\varepsilon\delta/2 \leq \int_{\delta/2}^{\delta} \sum_{i:Q_i\cap\gamma_r\neq\emptyset} h(Q_i)\,dr \leq \sum_{i\in I_{B(x,\delta)}} h(Q_i) \int_0^{\delta} \chi_{Q_i\cap\gamma_r}\,dr \qquad (2.6)$$

$$= \sum_{i\in I_{B(x,\delta)}} h(Q_i)d(Q_i),$$

where $d(Q_i) := \mathcal{H}^1(\{r \in [0,\delta] : Q_i \cap \gamma_r \neq \emptyset\})$. The Ahlfors regularity of the peripheral disks implies that there exists some uniform constant $K > 0$ such that $d(Q_i)^2 \leq K\mathcal{H}^2(Q_i \cap B(x, \delta))$; see Remark 2.3.5. Using the Cauchy–Schwarz inequality and this fact in (2.6) we obtain

$$\varepsilon^2\delta^2/4 \leq \sum_{i\in I_{B(x,\delta)}} h(Q_i)^2 \sum_{i\in I_{B(x,\delta)}} d(Q_i)^2$$

$$\leq K \sum_{i\in I_{B(x,\delta)}} h(Q_i)^2 \sum_{i\in\mathbb{N}} \mathcal{H}^2(Q_i \cap B(x, \delta))$$

$$= K \sum_{i\in I_{B(x,\delta)}} h(Q_i)^2 \cdot \mathcal{H}^2(B(x, \delta))$$

$$= C\delta^2 \sum_{i\in I_{B(x,\delta)}} h(Q_i)^2.$$

Hence, if δ is sufficiently small so that $\sum_{i \in I_{B(x,\delta)}} h(Q_i)^2 < \varepsilon^2/4C$ (see Remark 2.4.5) we obtain a contradiction.

In the case that $x \in \partial Q_{i_0}$ the same computations work if we exclude the index i_0 from the sums, since eventually we want to make $\sum_{i \in I_{B(x,\delta)} \setminus \{i_0\}} h(Q_i)^2$ arbitrarily small.

(b) Arguing as in part (a) we can find a small $\delta > 0$ such that the balls $B(x, \delta)$, $B(y, \delta)$ are disjoint, they are contained in Ω, and there exists a set $J \subset (\delta/2, \delta)$ of positive 1-measure such that

$$\sum_{\substack{i : Q_i \cap (\gamma_r^x \cup \gamma_r^y) \neq \emptyset \\ i \neq i_0}} h(Q_i) < \varepsilon$$

for all $r \in J$. We may also assume that for all $r \in J$ the paths γ_r^x, γ_r^y avoid the given set A with $\mathcal{H}^1(A) = 0$.

Let $\gamma_0 \subset\subset Q_{i_0}$ be a path that connects $\partial B(x, \delta/2) \cap Q_{i_0}$ to $\partial B(y, \delta/2) \cap Q_{i_0}$. Consider the function $\psi(\cdot) = \mathrm{dist}(\cdot, \gamma_0)$. Since $\mathrm{dist}(\gamma_0, \partial Q_{i_0}) > 0$, by Proposition 2.4.2 there exists $s_0 > 0$ such that for a.e. $s \in [0, s_0]$ there exists a path $\gamma_s \subset \psi^{-1}(s) \cap Q_{i_0}$ connecting $\partial B(x, \delta/2) \cap Q_{i_0}$ to $\partial B(y, \delta/2) \cap Q_{i_0}$. Then for all $r \in J$ and a.e. $s \in [0, s_0]$ the path γ_s connects the circular paths γ_r^x and γ_r^y. We claim that for a.e. $(r, s) \in J \times [0, s_0]$ all simple paths contained in the concatenation $\gamma_{r,s}$ of γ_r^x, γ_s, γ_r^y avoid a given path family Γ with $\mathrm{mod}_2(\Gamma) = 0$. Note here that $J \times [0, s_0]$ has positive 2-measure.

For each $\eta > 0$ we can find a function λ that is admissible for Γ with $\|\lambda\|_2 < \eta$. Let $T \subset J \times [0, s_0]$ be the set of (r, s) for which $\gamma_{r,s}$ has a simple subpath lying in Γ. Using the first co-area inequality in Proposition 2.4.1 we have

$$\mathcal{H}^2(T) = \int_T 1 \, d\mathcal{H}^2 \leq \int_J \int_0^{s_0} \left(\int_{\gamma_r^x} \lambda \, d\mathcal{H}^1 + \int_{\gamma_r^y} \lambda \, d\mathcal{H}^1 + \int_{\gamma_s} \lambda \, d\mathcal{H}^1 \right) ds \, dr$$

$$\leq C \|\lambda\|_2 \leq C\eta.$$

Letting $\eta \to 0$ we obtain that $\mathcal{H}^2(T) = 0$, as desired. This completes the proof of part (b). $\square$

Remark 2.4.8 The proof of part (a) shows the following stronger conclusion: there exists a constant $C > 0$ such that if $x \in S^\circ$, then there exists some $r \in [\delta/2, \delta]$ such that $\gamma_r(t) = x + re^{it}$ has the desired properties and

$$\sum_{i : Q_i \cap \gamma_r \neq \emptyset} h(Q_i) \leq C \left(\sum_{i \in I_{B(x,\delta)}} h(Q_i)^2 \right)^{1/2}.$$

Remark 2.4.9 The uniform Ahlfors 2-regularity of the peripheral disks Q_i was crucial in the proof. In fact, without the assumption of uniform Ahlfors 2-regularity, one can construct a relative Sierpiński carpet for which the conclusion of the lemma fails.

We also include a topological lemma:

Lemma 2.4.10 *The following statements are true:*

(a) *For each peripheral disk Q_{i_0}, there exists a Jordan curve $\gamma \subset S^\circ$ that contains Q_{i_0} in its interior and lies arbitrarily close to Q_{i_0} (in the Hausdorff sense). In particular, S° is dense in S.*

(b) *For any $x, y \in S$ there exists an open path $\gamma \subset S^\circ$ that joins x, y. Moreover, for each $r > 0$, if y is sufficiently close to x, the path γ can be taken so that $\gamma \subset B(x, r)$.*

The proof is an application of Moore's theorem [29] and can be found in [28, Proof of Theorem 5.2, p. 4331], in the case that Ω is a Jordan region. We include a proof of this more general statement here. We will use the following decomposition theorem, which is slightly stronger than Moore's theorem:

Theorem 2.4.11 ([15, Corollary 6A, p. 56]) *Let $\{Q_i\}_{i \in \mathbb{N}}$ be a sequence of Jordan regions in the sphere S^2 with disjoint closures and diameters converging to 0, and consider an open set $U \supset \bigcup_{i \in \mathbb{N}} \overline{Q}_i$. Then there exists a continuous, surjective map $f : S^2 \to S^2$ with the following properties: there exist countably many distinct points p_i, $i \in \mathbb{N}$, such that $f^{-1}(p_i) = \overline{Q}_i$ for $i \in \mathbb{N}$, and f is injective on $S^2 \setminus \bigcup_{i \in \mathbb{N}} \overline{Q}_i$ with $f(S^2 \setminus \bigcup_{i \in \mathbb{N}} \overline{Q}_i) = S^2 \setminus \{p_i : i \in \mathbb{N}\}$.*

Proof of Lemma 2.4.10 By Lemma 2.3.4 the spherical diameters of the peripheral disks Q_i converge to 0. Hence, we may apply the decomposition theorem with $U = \Omega$ and obtain the collapsing map $f : S^2 \to S^2$. A given peripheral disk $\overline{Q}_{i_0}$ is mapped to a point $p_{i_0} \in \Omega$. Arbitrarily close to p_{i_0} we can find round circles that avoid the countably many points that correspond to the collapsing of the other peripheral disks. The preimages of these round circles under f are Jordan curves $\gamma \subset S^2$ that are contained in Ω, lie in S°, and are contained in small neighborhoods of $\overline{Q}_{i_0}$. This completes the proof of part (a). For part (b), we consider three cases:

Case 1 Suppose first that $x, y \in S^\circ$. Using the decomposition theorem with $U = \Omega$, we obtain points $\tilde{x} = f(x)$ and $\tilde{y} = f(y)$ in $\Omega \setminus \{p_i : i \in \mathbb{N}\}$. We connect $\tilde{x}$ and $\tilde{y}$ with a path $\tilde{\alpha} \subset \Omega$ (recall that Ω is connected). By perturbing this path, we may obtain another path, still denoted by $\tilde{\alpha}$, that connects $\tilde{x}$ and $\tilde{y}$ in Ω but avoids the points $\{p_i : i \in \mathbb{N}\}$. Then $f^{-1}(\tilde{\alpha})$, by injectivity, yields the desired path in S° that connects x and y. In fact, if y is sufficiently close to x, the path $f^{-1}(\tilde{\alpha})$ can be taken to lie in a small neighborhood of x. To see this, first note that if y is close to x then $\tilde{y}$ is close to $\tilde{x}$ by the continuity of f. We can then find a path $\tilde{\alpha} \subset \Omega \setminus \{p_i : i \in \mathbb{N}\}$ connecting $\tilde{x}$ and $\tilde{y}$, and lying arbitrarily close to the line segment $[\tilde{x}, \tilde{y}]$. The lift $f^{-1}(\tilde{\alpha})$, by continuity, has to be contained in a small neighborhood of x.

If x or y lies on a peripheral circle, then we need to modify the preceding argument to obtain an open path $\gamma \subset S^\circ$ with endpoints x and y:

Case 2 Suppose that $x \in \partial Q_{i_0}$ and $y \in S^\circ$. Then in the application of the decomposition theorem we do not collapse the peripheral disk $\overline{Q}_{i_0}$. We set $U = \Omega \cup Q_{i_0}$ and we collapse $\{\overline{Q}_i : i \in \mathbb{N} \setminus \{i_0\}\}$ to points with a map $f : S^2 \to S^2$ that is the identity outside U, as in the statement of the decomposition theorem. Then we consider an open path $\tilde{\alpha} \subset \Omega \setminus \overline{Q}_{i_0}$ connecting $\tilde{x}$ and $\tilde{y}$; in order to find such a path one can assume that $\overline{Q}_{i_0}$ is a round disk by using a homeomorphism of S^2. Now, the path $\tilde{\alpha}$ can be modified to avoid the countably many points corresponding to the collapsed peripheral disks. This modified path lifts under f to the desired path, and as before, it can be taken to lie arbitrarily close to x, provided that $y \in S^\circ$ is sufficiently close to x. This proves (b) in this case.

Case 3 Finally, suppose that $x \in \partial Q_{i_0}$ and $y \in \partial Q_i$ for some $i_0, i \in \mathbb{N}$. Then we can connect the points x, y to points x', $y' \in S^\circ$ with open paths $\gamma_x, \gamma_y \subset S^\circ$ by Case 2. Case 1 implies that the points x', y' can be connected with a path $\gamma \subset S^\circ$. Concatenating γ_x, γ, and γ_y yields an open path in S° that connects x and y.

It remains to show that if y is sufficiently close to x, then there exists an open path $\gamma \subset S^\circ$ connecting x and y that lies near x. By the density of S° in S (from part (a)) we may find arbitrarily close to y points $z \in S^\circ$. By Case 2, for each $r > 0$ there exists $\delta > 0$ such that if $|z - x| < \delta$ and $z \in S^\circ$, then there exists an open path $\gamma \subset B(x, r) \cap S^\circ$ that connects x and z. We assume that $|y - x| < \delta/2$ and $y \in \partial Q_i$ for some $i \in \mathbb{N}$. Then using the conclusion of Case 2 we consider a point $z \in S^\circ \in B(x, \delta)$ close to y and an open path $\gamma_1 \subset B(x, r) \cap S^\circ$ that connects y to z. Since $z \in B(x, \delta)$, there exists another open path $\gamma_2 \subset B(x, r) \cap S^\circ$ connecting z to x. Concatenating γ_1 and γ_2 provides the desired path. □

Finally, we include a technical lemma:

Lemma 2.4.12 *Suppose that $\gamma \subset \mathbb{C}$ is a non-constant path with $\mathcal{H}^1(\gamma \cap S) = 0$.*

(a) *If $x \in \gamma \cap S^\circ$, then arbitrarily close to x (in the Hausdorff sense) we can find peripheral disks Q_i with $Q_i \cap \gamma \neq \emptyset$.*

(b) *If γ is an open path that does not intersect a peripheral disk Q_{i_0} for some $i_0 \in \mathbb{N}$, and $x \in \overline{\gamma} \cap \partial Q_{i_0}$, then arbitrarily close to x we can find peripheral disks Q_i, $i \neq i_0$, with $Q_i \cap \gamma \neq \emptyset$.*

Proof If either of the two statements failed, then there would exist a small ball $B(x, \varepsilon)$, not containing γ, such that all points $y \in \gamma \cap B(x, \varepsilon)$ lie in S. Since γ is connected and it exits $B(x, \varepsilon)$, there exists a continuum $\beta \subset \gamma \cap B(x, \varepsilon) \cap S$ with $\operatorname{diam}(\beta) \geq \varepsilon/2$. Then $\mathcal{H}^1(\gamma \cap S) \geq \operatorname{diam}(\beta) > 0$, a contradiction. □

2.5 Sobolev Spaces on Relative Sierpiński Carpets

In this section we treat one of the main objects of the chapter, the Sobolev spaces on relative Sierpiński carpets. This is the class of maps among which we would like to minimize a type of *Dirichlet energy*, in order to define carpet-harmonic functions.

We will start with a preliminary version of our Sobolev functions, namely with *discrete Sobolev functions*. These are only defined on the set peripheral disks, i.e., the "holes" of the carpet. Using them, we define a non-discrete version of Sobolev functions whose domain is the carpet S. This will be the Sobolev space that we will use in our developments. As it turns out, the two spaces of functions are just two sides of the same coin, since they are isomorphic. The reasons for introducing the discrete Sobolev functions are because, in practice, the definition is much easier to check, and limiting theorems are proved with less effort.

2.5.1 Discrete Sobolev Spaces

Let $\hat{f} \colon \{Q_i\}_{i \in \mathbb{N}} \to \mathbb{R}$ be a map defined on the set of peripheral disks $\{Q_i\}_{i \in \mathbb{N}}$. We say that the sequence $\{\rho(Q_i)\}_{i \in \mathbb{N}}$ is a *weak* (resp. *strong*) *upper gradient* for $\hat{f}$ if there exists an exceptional family Γ_0 of paths in Ω with $\text{mod}_w(\Gamma_0) = 0$ (resp. $\text{mod}_s(\Gamma_0) = 0$) such that for all paths $\gamma \subset \Omega$ with $\gamma \notin \Gamma_0$ and all peripheral disks Q_{i_j} with $Q_{i_j} \cap \gamma \neq \emptyset$, $j = 1, 2$, we have

$$|\hat{f}(Q_{i_1}) - \hat{f}(Q_{i_2})| \leq \sum_{i:Q_i \cap \gamma \neq \emptyset} \rho(Q_i). \qquad (2.7)$$

Using the upper gradients we define:

Definition 2.5.1 Let $\hat{f} \colon \{Q_i\}_{i \in \mathbb{N}} \to \mathbb{R}$ be a map defined on the set of peripheral disks $\{Q_i\}_{i \in \mathbb{N}}$. We say that $\hat{f}$ lies in the *local weak* (resp. *strong*) *Sobolev space* $\widehat{W}^{1,2}_{w,\text{loc}}(S)$ (resp. $\widehat{W}^{1,2}_{s,\text{loc}}(S)$) if there exists a weak (resp. strong) upper gradient $\{\rho(Q_i)\}_{i \in \mathbb{N}}$ for $\hat{f}$ such that for every ball $B \subset\subset \Omega$ we have

$$\sum_{i \in I_B} \hat{f}(Q_i)^2 \, \text{diam}(Q_i)^2 < \infty, \quad \text{and} \qquad (2.8)$$

$$\sum_{i \in I_B} \rho(Q_i)^2 < \infty. \qquad (2.9)$$

Furthermore, if these conditions hold for the full sums over $i \in \mathbb{N}$, then we say that $\hat{f}$ lies in the *weak* (resp. *strong*) *Sobolev space* $\widehat{W}^{1,2}_w(S)$ (resp. $\widehat{W}^{1,2}_s(S)$).

In order to avoid making every time the distinction between weak and strong Sobolev spaces, we will also use the notation $\widehat{W}^{1,2}_{*,\mathrm{loc}}$, $\widehat{W}^{1,2}_{*}$, where $*$ can mean either "w" or "s". For example, the statement $\widehat{W}^{1,2}_{*} \subset \widehat{W}^{1,2}_{*,\mathrm{loc}}$ means that $\widehat{W}^{1,2}_{w} \subset \widehat{W}^{1,2}_{w,\mathrm{loc}}$ and $\widehat{W}^{1,2}_{s} \subset \widehat{W}^{1,2}_{s,\mathrm{loc}}$.

Remark 2.5.2 Conditions (2.8) and (2.9) are equivalent to saying that the sequences $\{\hat{f}(Q_i)\,\mathrm{diam}(Q_i)\}_{i\in\mathbb{N}}$ and $\{\rho(Q_i)\}_{i\in\mathbb{N}}$ are locally square-summable. Also, if, e.g., $\hat{f}(Q_i)$ is bounded then (2.8) holds since we have $\sum_{i\in I_B} \mathrm{diam}(Q_i)^2 < \infty$ for all $B \subset\subset \Omega$, by Corollary 2.3.6.

Remark 2.5.3 The classical Sobolev spaces $W^{1,2}_{\mathrm{loc}}(\mathbb{R}^n)$ and $W^{1,2}(\mathbb{R}^n)$ can be defined using a suitable notion of weak upper gradients, in the same spirit as the above definitions. In fact, our definition is motivated by the *Newtonian spaces* $N^{1,p}(\mathbb{R}^n)$, which contain *good* representatives of Sobolev functions, rather than equivalence classes of functions; see [22] for background on weak upper gradients and Newtonian spaces.

If $\mathrm{mod}_s(\Gamma_0) = 0$, then $\mathrm{mod}_w(\Gamma_0) = 0$ by Lemma 2.3.1. This shows that if $\{\rho(Q_i)\}_{i\in\mathbb{N}}$ is a strong upper gradient for $\hat{f}$, then it is also a weak upper gradient for $\hat{f}$. Thus,

$$\widehat{W}^{1,2}_{s,\mathrm{loc}}(S) \subset \widehat{W}^{1,2}_{w,\mathrm{loc}}(S). \tag{2.10}$$

Remark 2.5.4 We have not been able to show the reverse inclusion, which probably depends on the geometry of the peripheral disks and their separation. A conjecture could be that the two spaces agree if the peripheral circles of the carpet are uniform quasicircles and they are *uniformly relatively separated*, i.e., there exists a uniform constant $\delta > 0$ such that the *relative distance*

$$\Delta(Q_i, Q_j) = \frac{\mathrm{dist}(Q_i, Q_j)}{\min\{\mathrm{diam}(Q_i), \mathrm{diam}(Q_j)\}}$$

satisfies $\Delta(Q_i, Q_j) \geq \delta$ for all $i \neq j$.

For the rest of the section we fix a function $\hat{f} \in \widehat{W}^{1,2}_{*,\mathrm{loc}}(S)$. Our goal is to construct a function f that is defined on the entire carpet S and is a regularized version of the discrete function $\hat{f}$. Let $\mathcal{G}$ be the family of *good paths* γ in Ω with

(1) $\mathcal{H}^1(\gamma \cap S) = 0$,
(2) the upper gradient inequality (2.7) is satisfied for all subpaths of γ,
(3) $\sum_{i:Q_i\cap\gamma'\neq\emptyset} \rho(Q_i) < \infty$ for all subpaths γ' of γ which are compactly contained in Ω,
(3*) in the case $\sum_{i\in\mathbb{N}} \rho(Q_i)^2 < \infty$, we require $\sum_{i:Q_i\cap\gamma\neq\emptyset} \rho(Q_i) < \infty$.

It is immediate to see that all subpaths of a path $\gamma \in \mathcal{G}$ also lie in $\mathcal{G}$. Note that $\mathcal{G}$ depends both on $\{\hat{f}(Q_i)\}_{i\in\mathbb{N}}$ and on $\{\rho(Q_i)\}_{i\in\mathbb{N}}$. Property (1) is crucial and will allow us to apply Lemma 2.4.12.

Lemma 2.5.5 *Suppose that* $\hat{f} \in \widehat{W}^{1,2}_{w,\text{loc}}(S)$ *(resp.* $\widehat{W}^{1,2}_{s,\text{loc}}(S)$*). Then the comple-ment of* $\mathcal{G}$ *(i.e., all curves in* Ω *that do not lie in* $\mathcal{G}$*) has weak (resp. strong) modulus equal to* 0.

In other words, $\mathcal{G}$ contains almost every path $\gamma \subset \Omega$.

Proof By the subadditivity of modulus, it suffices to show that the family of curves for which one of the conditions (1),(2),(3), or (3*) is violated has weak (resp. strong) modulus equal to 0.

We first note that $\mathcal{H}^1(\gamma \cap S) = 0$ holds for all paths outside a family of strong carpet modulus equal to 0, so the weak carpet modulus is also equal to 0 by Lemma 2.3.1.

Moreover, Remark 2.4.6 implies that the family of paths that have a subpath lying in a family of weak (resp. strong) carpet modulus equal to 0 has itself weak (resp. strong) carpet modulus zero. This justifies that the upper gradient inequality holds for *all* subpaths of paths γ that lie outside an exceptional family of weak (resp. strong) modulus zero.

Finally, since for any set $V \subset\subset \Omega$ we have $\sum_{i \in I_V} \rho(Q_i)^2 < \infty$ by (2.9), it follows that ρ (as well as any positive multiple of it) is an admissible weight for the strong modulus of the family of paths in Ω that have a subpath γ' in V with $\sum_{i:Q_i \cap \gamma' \neq \emptyset} \rho(Q_i) = \infty$. It follows that this path family has strong (and thus weak) carpet modulus equal to zero. Exhausting Ω by compact sets V and using the subadditivity of modulus, we obtain that the family of paths in Ω that have a subpath compactly contained in Ω, with $\sum_{i:Q_i \cap \gamma' \neq \emptyset} \rho(Q_i) = \infty$, has strong (and thus weak) carpet modulus equal to zero. The case (3*) has the same proof, since $\sum_{i \in \mathbb{N}} \rho(Q_i)^2 < \infty$ there.   □

We say that a point $x \in S$ is *accessible by a curve* $\gamma_0 \in \mathcal{G}$, if $x \in \gamma_0 \cap S^\circ$, or if there exists an open subcurve γ of γ_0 such that $x \notin \gamma$, $x \in \overline{\gamma}$, and γ does not meet the (interior of the) peripheral disk Q_{i_0} whenever $x \in \partial Q_{i_0}$; see Fig. 2.1. In the first case we set $\gamma = \gamma_0$. By Lemma 2.4.12, if x is accessible by γ_0, then arbitrarily close to x we can find peripheral disks Q_i with $Q_i \cap \gamma \neq \emptyset$. We say that a point is *accessible* if it is accessible by some path in the implicitly understood family of paths. For a point $x \in S$ that is accessible by γ_0 we define

$$f(x) = \liminf_{\substack{Q_i \to x \\ Q_i \cap \gamma \neq \emptyset}} \hat{f}(Q_i), \tag{2.11}$$

where γ is a subpath of γ_0 as above. Often, we will abuse terminology and we will use the above definition if x is accessible by γ, without mentioning that γ is a subpath of γ_0; see e.g. the statement of Lemma 2.5.6. Using Lemma 2.4.3 it is easy to construct non-exceptional paths in $\mathcal{G}$ passing through a given continuum. In fact, it can be shown that for each peripheral circle ∂Q_{i_0} there is a dense set of points which are accessible by good paths $\gamma \in \mathcal{G}$. For non-accessible points $x \in \partial Q_{i_0}$ we define $f(x)$ as $\liminf f(y)$, as y approaches x through accessible points $y \in \partial Q_{i_0}$.

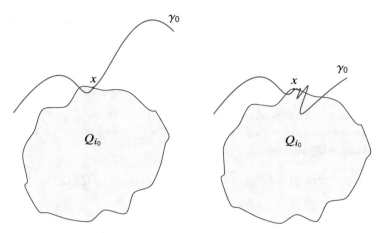

Fig. 2.1 A point $x \in \partial Q_{i_0}$ that is accessible by γ_0 (left) and non-accessible (right)

For the other points of the carpet S (which must belong to S°) we define $f(x)$ as $\liminf f(y)$, as y approaches x through accessible points.

First we show that the map $f \colon S \to \widehat{\mathbb{R}}$ is well-defined, i.e., the definition does not depend on the curve γ from which the point x is accessible.

Lemma 2.5.6 *If $\gamma_1, \gamma_2 \in \mathcal{G}$ are two paths from which the point $x \in S$ is accessible, then*

$$\liminf_{\substack{Q_i \to x \\ Q_i \cap \gamma_1 \neq \emptyset}} \hat{f}(Q_i) = \liminf_{\substack{Q_i \to x \\ Q_i \cap \gamma_2 \neq \emptyset}} \hat{f}(Q_i).$$

As remarked before, here *accessible by γ_1* means that there exists an open subpath of γ_1 that we still denote by γ_1 such that $x \in \overline{\gamma}_1 \setminus \gamma_1$, and γ_1 does not intersect the peripheral disk Q_{i_0}, in the case $x \in \partial Q_{i_0}$.

Proof We may assume that γ_1, γ_2 are compactly contained in Ω, otherwise we consider subpaths of them. We fix $\varepsilon > 0$ and consider peripheral disks $Q_{i_j}, j = 1, 2$, very close to x such that $Q_{i_j} \cap \gamma_j \neq \emptyset$ (the existence of Q_{i_j} can be justified by Lemma 2.4.12(b), since $\mathcal{H}^1(\gamma_j \cap S) = 0$) and such that the truncated paths $\gamma'_j \subset \gamma_j$ that join Q_{i_j} to x have short ρ-length, i.e.,

$$\sum_{i \colon Q_i \cap \gamma'_j \neq \emptyset} \rho(Q_i) < \varepsilon \tag{2.12}$$

for $j = 1, 2$. This can be done since each γ_j is compactly contained in Ω and $\gamma_j \in \mathcal{G}$, which implies that $\sum_{i \colon Q_i \cap \gamma_j \neq \emptyset} \rho(Q_i) < \infty$ for $j = 1, 2$.

We first assume that $x \in S^\circ$. Using Lemma 2.4.7(a), we can find a circular path $\gamma_r(t) = x + re^{it}$ lying in $\mathcal{G}$ with r smaller than the distance between Q_{i_j} and x

for $j = 1, 2$, such that γ_r does not intersect the sets $\gamma_j' \cap S$, which have Hausdorff 1-measure zero, and

$$\sum_{i:Q_i \cap \gamma_r \neq \emptyset} \rho(Q_i) < \varepsilon. \tag{2.13}$$

Note that the radius r depends on Q_{i_j}, $j = 1, 2$. It follows that γ_r has to meet some Q_{i_j}' that is intersected by γ_j', $j = 1, 2$, so by the triangle inequality and the upper gradient inequality we have

$$|\hat{f}(Q_{i_1}) - \hat{f}(Q_{i_2})| \leq |\hat{f}(Q_{i_1}) - \hat{f}(Q_{i_1}')| + |\hat{f}(Q_{i_1}') - \hat{f}(Q_{i_2}')|$$
$$+ |\hat{f}(Q_{i_2}') - \hat{f}(Q_{i_2})|$$
$$\leq \sum_{i:Q_i \cap \gamma_1' \neq \emptyset} \rho(Q_i) + \sum_{i:Q_i \cap \gamma_r \neq \emptyset} \rho(Q_i) + \sum_{i:Q_i \cap \gamma_2' \neq \emptyset} \rho(Q_i)$$
$$< 3\varepsilon$$

by (2.12) and (2.13). The conclusion follows in this case.

In the case that $x \in \partial Q_{i_0}$ for some i_0, the only modification we have to make is in the application of Lemma 2.4.7(a), which now yields a path γ_r for sufficiently small r such that

$$\sum_{\substack{i:Q_i \cap \gamma_r \neq \emptyset \\ i \neq i_0}} \rho(Q_i) < \varepsilon.$$

Since the curves γ_1, γ_2 have x as an endpoint but they do not intersect Q_{i_0}, there exists a subarc γ_r' of γ_r that defines a *crosscut* separating ∞ from x in $\mathbb{R}^2 \setminus Q_{i_0}$ and having the following properties (see Fig. 2.2): γ_r' meets peripheral disks Q_{i_j}' that are also intersected by γ_j', $j = 1, 2$, $\gamma_r' \cap Q_{i_0} = \emptyset$ but the endpoints of γ_r' lie on ∂Q_{i_0}, and

$$\sum_{i:Q_i \cap \gamma_r' \neq \emptyset} \rho(Q_i) < \varepsilon.$$

We direct the reader to [31, Chap. 2.4] for a discussion on crosscuts. If we use this in the place of (2.13), the argument in the previous paragraph yields the conclusion. Here, we remark that γ_r' is a good path lying in G, since it is a subpath of $\gamma_r \in G$; see Lemma 2.5.5.   □

For a function $g: S \to \widehat{\mathbb{R}}$ and a peripheral disk Q_i we define

$$M_{Q_i}(g) = \sup_{x \in \partial Q_i} g(x) \quad \text{and} \quad m_{Q_i}(g) = \inf_{x \in \partial Q_i} g(x).$$

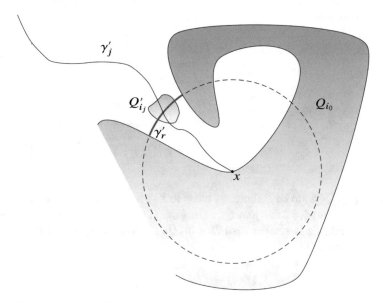

Fig. 2.2 The crosscut defined by γ_r

If at least one of the above quantities is finite, then we define

$$\operatorname*{osc}_{Q_i}(g) = M_{Q_i}(g) - m_{Q_i}(g).$$

Corollary 2.5.7 *For each peripheral disk Q_{i_0} the quantities $M_{Q_{i_0}}(f)$, $m_{Q_{i_0}}(f)$, and $\operatorname{osc}_{Q_{i_0}}(f)$ are finite, and in particular,*

$$\operatorname*{osc}_{Q_{i_0}}(f) \leq \rho(Q_{i_0}) \quad \text{and} \quad |M_{Q_{i_0}}(f) - \hat{f}(Q_{i_0})| \leq \rho(Q_{i_0}).$$

Proof Let $x, y \in \partial Q_{i_0}$ be points that are accessible by γ_x, γ_y, respectively, and let $Q_{i_x}, Q_{i_y}, i_x, i_y \neq i_0$, be peripheral disks intersected by γ_x, γ_y, close to x, y, respectively. Applying Lemma 2.4.7(b) for small $\varepsilon > 0$, we obtain that there exists a small $r < \min\{\operatorname{dist}(x, Q_{i_x}), \operatorname{dist}(y, Q_{i_y})\}$ such that the concatenation γ of the circular paths γ_r^x, γ_r^y with a path $\gamma_0 \subset Q_{i_0}$ intersects peripheral disks Q'_{i_x}, Q'_{i_y} on γ_x, γ_y, respectively, and

$$\sum_{i:Q_i \cap \gamma \neq \emptyset} \rho(Q_i) = \sum_{\substack{i:Q_i \cap \gamma \neq \emptyset \\ i \neq i_0}} \rho(Q_i) + \rho(Q_{i_0}) < \varepsilon + \rho(Q_{i_0}).$$

We replace γ by a simple subpath of it connecting Q'_{i_x} with Q'_{i_y}, and we still denote this subpath by γ. By Lemma 2.4.7(b) γ is a good path in $\mathcal{G}$, so by the upper gradient inequality we have

$$|\hat{f}(Q_{i_x}) - \hat{f}(Q_{i_y})| \leq |\hat{f}(Q_{i_x}) - \hat{f}(Q'_{i_x})| + |\hat{f}(Q'_{i_x}) - \hat{f}(Q'_{i_y})|$$

$$+ |\hat{f}(Q'_{i_y}) - \hat{f}(Q_{i_y})|$$

$$\leq \sum_{i: Q_i \cap \gamma_x \neq \emptyset} \rho(Q_i) + \sum_{i: Q_i \cap \gamma \neq \emptyset} \rho(Q_i) + \sum_{i: Q_i \cap \gamma_y \neq \emptyset} \rho(Q_i)$$

$$\leq \sum_{i: Q_i \cap \gamma_x \neq \emptyset} \rho(Q_i) + \rho(Q_{i_0}) + \varepsilon + \sum_{i: Q_i \cap \gamma_y \neq \emptyset} \rho(Q_i).$$

By taking Q_{i_x}, Q_{i_y} to be sufficiently close to x, y and also, by truncating γ_x, γ_y to smaller subpaths γ'_x, γ'_y that join Q_{i_x}, Q_{i_y} to x, y, respectively, we may assume that the right hand side is smaller than $2\varepsilon + \rho(Q_{i_0})$. Thus, taking limits as $Q_{i_x} \to x$, $Q_{i_y} \to y$, we obtain

$$|f(x) - f(y)| \leq \rho(Q_{i_0}). \tag{2.14}$$

One can use Lemma 2.4.7(a) and argue in the same way to derive that for all accessible points $x \in \partial Q_{i_0}$ we have

$$|f(x) - \hat{f}(Q_{i_0})| \leq \rho(Q_{i_0}). \tag{2.15}$$

This shows that in fact $f(x)$ is uniformly bounded for accessible points $x \in \partial Q_{i_0}$. Since accessible points are dense in ∂Q_{i_0}, the bounds in (2.14) and (2.15) also hold on non-accessible points x, y of the peripheral circle ∂Q_{i_0}, by the definition of f. This completes the proof.   □

Before proceeding we define:

Definition 2.5.8 Let $g: S \to \widehat{\mathbb{R}}$ be an extended function. We say that $\{\rho(Q_i)\}_{i \in \mathbb{N}}$ is a *weak* (resp. *strong*) *upper gradient* for g if there exists an exceptional family Γ_0 of paths in Ω with $\mathrm{mod}_w(\Gamma_0) = 0$ (resp. $\mathrm{mod}_s(\Gamma_0) = 0$) such that for all paths $\gamma \subset \Omega$ with $\gamma \notin \Gamma_0$ and all points $x, y \in \gamma \cap S$ we have $g(x), g(y) \neq \pm\infty$ and

$$|g(x) - g(y)| \leq \sum_{i: Q_i \cap \gamma \neq \emptyset} \rho(Q_i).$$

If it is implicitly understood whether an upper gradient is weak or strong, we will only refer to it as an upper gradient.

The map f that we constructed inherits the properties of $\hat{f}$ in the Sobolev space definition. Namely:

Proposition 2.5.9 *The sequence $\{\mathrm{osc}_{Q_i}(f)\}_{i\in\mathbb{N}}$ is a upper gradient for f. Moreover, for every ball $B \subset\subset \Omega$ we have*

$$\sum_{i\in I_B} M_{Q_i}(f)^2 \,\mathrm{diam}(Q_i)^2 < \infty, \text{ and}$$

$$\sum_{i\in I_B} \mathrm{osc}_{Q_i}(f)^2 < \infty.$$

The proof uses the following topological lemma, which we prove later.

Lemma 2.5.10 *Let $\gamma \subset \Omega$ be a path in $\mathcal{G}$, and let $J \subset \mathbb{N}$ be a finite index set. Assume that γ has endpoints $x, y \in S$, but γ does not intersect any peripheral disk Q_i with $x \in \partial Q_i$ or $y \in \partial Q_i$. Then there exist finitely many subpaths $\gamma_1, \ldots, \gamma_m$ of γ having endpoints in S with the following properties:*

  (i) *$\gamma_i \in \mathcal{G}$ for all $i \in \{1, \ldots, m\}$,*
 (ii) *γ_i intersects only peripheral disks that are intersected by γ, for all $i \in \{1, \ldots, m\}$,*
(iii) *γ_i and γ_j intersect disjoint sets of peripheral disks for $i \neq j$,*
 (iv) *γ_i does not intersect peripheral disks Q_j, $j \in J$, for all $i \in \{1, \ldots, m\}$,*
  (v) *γ_1 starts at $x_1 = x$, γ_m terminates at $y_m = y$, and in general the path γ_i starts at x_i and terminates at y_i such that for each $i \in \{1, \ldots, m-1\}$ we either have*

  - *$y_i = x_{i+1}$, i.e., γ_i and γ_{i+1} have a common endpoint, or*
  - *$y_i, x_{i+1} \in \partial Q_{j_i}$ for some $j_i \in \mathbb{N}$, i.e., γ_i and γ_{i+1} have an endpoint on some peripheral circle ∂Q_{j_i}.*

  The peripheral disks Q_{j_i} that arise from the second case are distinct and they are all intersected by the original curve γ.

Note that properties (i) and (ii) hold automatically for subpaths of γ, so (iii),(iv), and (v) are the most crucial properties.

Proof of Proposition 2.5.9 Note that the latter two claims follow immediately from Corollary 2.5.7, if we observe that $\mathrm{diam}(Q_i)$ is bounded for $i \in I_B$, where B is a ball compactly contained in Ω. The latter is incorporated in the definition of a relative Sierpiński carpet; see Sect. 2.2.

For the first claim, we will show that for every $\gamma \in \mathcal{G}$ and points $x, y \in \gamma \cap S$ we have

$$|f(x) - f(y)| \le \sum_{i: Q_i \cap \gamma \neq \emptyset} \mathrm{osc}_{Q_i}(f).$$

Fix a good path $\gamma \in \mathcal{G}$ that has x, y as its endpoints, and assume it is parametrized so that it travels from x to y. If $x \in \partial Q_{i_x}$, let $x' \in \partial Q_{i_x}$ be the point of last exit of γ

from Q_{i_x}. Similarly, consider the point $y' \in \partial Q_{i_y}$ of first entry of γ in Q_{i_y} (after x'), in the case $y \in \partial Q_{i_y}$. Observe that the points x' and y', if defined, are accessible by γ. Since $|f(x) - f(x')| \leq \mathrm{osc}_{Q_{i_x}}(f) \leq \rho(Q_{i_x})$ and $|f(y) - f(y')| \leq \mathrm{osc}_{Q_{i_y}}(f) \leq \rho(Q_{i_y})$, it suffices instead to prove the statement for x', y' and the subpath of γ connecting them as described.

In particular, we assume that $\gamma \in \mathcal{G}$ is a good path that connects x, y but it does not intersect ∂Q_{i_x} or ∂Q_{i_y} in the case $x \in \partial Q_{i_x}$ or $y \in \partial Q_{i_y}$. Hence, the path γ can be used to define both $f(x)$ and $f(y)$ by (2.11). If Q_{i_1}, Q_{i_2} are peripheral disks intersected by γ close to x, y, respectively, then by the upper gradient inequality for $\hat{f}$ we have

$$|\hat{f}(Q_{i_1}) - \hat{f}(Q_{i_2})| \leq \sum_{i: Q_i \cap \gamma \neq \emptyset} \rho(Q_i).$$

Taking limits as $Q_{i_1} \to x$ and $Q_{i_2} \to y$ along γ (which also shows that $f(x), f(y) \neq \pm\infty$), we have

$$|f(x) - f(y)| \leq \sum_{i: Q_i \cap \gamma \neq \emptyset} \rho(Q_i). \tag{2.16}$$

By our preceding remarks, this also holds if γ intersects peripheral disks Q_i with $x \in \partial Q_i$ or $y \in \partial Q_i$. However, we would like to prove this statement with $\mathrm{osc}_{Q_i}(f)$ in the place of $\rho(Q_i)$.

Since $\gamma \in \mathcal{G}$ and it is compactly contained in Ω, we have $\sum_{i: Q_i \cap \gamma \neq \emptyset} \rho(Q_i) < \infty$ so for fixed $\varepsilon > 0$ there exists a finite index set $J \subset \mathbb{N}$ such that

$$\sum_{\substack{i: Q_i \cap \gamma \neq \emptyset \\ i \in \mathbb{N} \setminus J}} \rho(Q_i) < \varepsilon. \tag{2.17}$$

We consider curves $\gamma_1, \ldots, \gamma_m$ as in Lemma 2.5.10 with their endpoints, as denoted in Lemma 2.5.10. If the points y_k, x_{k+1} lie on the same peripheral circle ∂Q_{j_k}, we have $|f(y_k) - f(x_{k+1})| \leq \mathrm{osc}_{Q_{j_k}}(f)$. Otherwise, by the first alternative in (v), we have $y_k = x_{k+1}$, so $|f(y_k) - f(x_{k+1})| = 0$. Also, note that the curves γ_k lie in $\mathcal{G}$ (by (i)) and (2.16) holds for their endpoints. Therefore,

$$|f(x) - f(y)| \leq \sum_{k=1}^{m} |f(y_k) - f(x_k)| + \sum_{k=1}^{m-1} |f(y_k) - f(x_{k+1})|$$

$$\leq \sum_{k=1}^{m} \sum_{i: Q_i \cap \gamma_k \neq \emptyset} \rho(Q_i) + \sum_{j_k} \mathrm{osc}_{Q_{j_k}}(f).$$

Using (iii) and (iv), we see that the curves γ_k intersect disjoint sets of peripheral disks Q_j, $j \notin J$, which are all intersected by γ (property (ii)). Thus, the first term can be bounded by the expression in (2.17), and hence by ε. The second term is just bounded by the full sum of $\mathrm{osc}_{Q_i}(f)$ over γ (since the peripheral disks Q_{j_k} are distinct and γ intersects them by (v)), hence we obtain

$$|f(x) - f(y)| \leq \varepsilon + \sum_{i:Q_i \cap \gamma \neq \emptyset} \underset{Q_i}{\mathrm{osc}}(f).$$

Letting $\varepsilon \to 0$ yields the result. $\qquad\square$

Now, we move to the proof of Lemma 2.5.10.

Proof of Lemma 2.5.10 By ignoring some indices of J, we may assume that $Q_j \cap \gamma \neq \emptyset$ for all $j \in J$. The idea is to consider subpaths of γ joining peripheral circles ∂Q_j, $j \in J$, without intersecting Q_j, $j \in J$, and truncate them whenever they intersect some common peripheral disk.

More precisely, we assume that γ is parametrized as it runs from x to y we let $\tilde{\gamma}_1$ be the subpath of γ from $x = x_1$ until the first entry point of γ into the first peripheral disk Q_{i_1}, $i_1 \in J$, that γ meets, among the peripheral disks Q_j, $j \in J$. We let $x_2 \in \partial Q_{i_1}$ be the point of last exit of γ from ∂Q_{i_1} as it travels towards y. Now, we repeat the procedure with x_1 replaced by x_2, γ replaced by the subpath of γ from x_2 to y, and J replaced by $J \setminus \{i_1\}$. The procedure will terminate in the m-th step if the subpath of γ from x_m to y does not intersect any peripheral disk Q_j, $j \in J \setminus \{i_1, \ldots, i_{m-1}\}$. This will be the path $\tilde{\gamma}_m$. Note that the indices $i_1, \ldots, i_{m-1}$ are distinct.

By construction, the paths $\tilde{\gamma}_1, \ldots, \tilde{\gamma}_m$ do not intersect peripheral disks Q_j, $j \in J$, but they might still intersect common peripheral disks. Thus, we might need to truncate some of them in order to obtain paths with the desired properties. We do this using the following claim, which that we prove later:

Claim There exist closed paths $\gamma_1', \ldots, \gamma_{m'}'$, where $m' \leq m$, such that γ_i' is a subpath of some $\tilde{\gamma}_j$ and the following hold:

(a) γ_1' starts at the starting point $x_1 = x$ of γ_1, $\gamma_{m'}'$ terminates at the terminating point $y_m = y$ of γ_m,
(b) for each $i \in \{1, \ldots, m' - 1\}$ the path γ_i' starts at x_i' and terminates at y_i' such that we either have

(b1) $y_i' = x_{i+1}'$, i.e., γ_i' and γ_{i+1}' have a common endpoint, or
(b2) there exists $j_i \in \{1, \ldots, m - 1\}$ such that $y_i' = y_{j_i}$ and $x_{i+1}' = x_{j_i+1}$, i.e., γ_i' and γ_{i+1}' have a common endpoint with $\tilde{\gamma}_{j_i}$ and $\tilde{\gamma}_{j_i+1}$, respectively.

The indices j_i arising from the second case are distinct.

Moreover, the closed paths γ_i' are disjoint, with the exceptions of some constant paths and of the case (b1) in which "consecutive" paths can share only one endpoint.

Essentially, the conclusion of the claim is that we can obtain a family of disjoint (except at their endpoints) subpaths of $\tilde{\gamma}_1, \ldots, \tilde{\gamma}_m$ that have the same properties as $\tilde{\gamma}_1, \ldots, \tilde{\gamma}_m$.

Note that the paths γ_i' do not intersect peripheral disks Q_j, $j \in J$, since they are subpaths of paths $\tilde{\gamma}_j$. Also, whenever (b2) occurs, the paths γ_i' and γ_{i+1}' have an endpoint on some peripheral circle ∂Q_{j_i}, $j_i \in J$, and the indices j_i are distinct, by the construction of the curves $\tilde{\gamma}_i$. We discard the constant paths from the collection γ_i'. Then, after re-enumerating, we still have the preceding statement. It only remains to shrink the paths γ_i' such that property (iii) in the statement of Lemma 2.5.10 holds. For simplicity we denote the endpoints of the paths γ_i' by x_i and y_i. By replacing γ_i' with a subpath, we may assume that γ_i' does not return to x_i or y_i twice; in other words, if we parametrize $\gamma_i' : [0, 1] \to \mathbb{C}$, then $\gamma_i'(t) \neq x_i, y_i$ for all $t \in (0, 1)$.

By the definition of a relative Sierpiński carpet, the peripheral disks staying in a compact subset of Ω have diameters shrinking to zero. Hence, any point of γ_i' that has positive distance from a curve γ_j', $j \neq i$, has an open neighborhood with the property that it only intersects finitely many peripheral disks among the peripheral disks that intersect both γ_i' and γ_j'.

This observation implies that if we have two paths γ_i' and γ_j', $j > i$, that intersect some common peripheral disks, then we can talk about the "first" such peripheral disk Q_{i_0} that γ_i' meets as in travels from x_i to y_i. It is crucial here that γ_i' is disjoint from γ_j', except possibly for the endpoints y_i and x_j which could agree if $j = i + 1$ and we are in the case (a) of the Claim; in particular, x_i has positive distance from γ_j' and y_j has positive distance from γ_i'.

Another important observation from our Claim is that for each i, each of the endpoints of γ_i' is either an endpoint of some $\tilde{\gamma}_{j_i}$, and therefore lies on the boundary of some peripheral disk Q_{j_i} that is intersected by γ, or it is a common endpoint of γ_i' and γ_{i+1}'. In the latter case, if this endpoint does not lie in S, then it lies in some peripheral disk Q_{i_0} that is intersected by both curves γ_i' and γ_{i+1}'. The truncating procedure explained below ensures that the paths γ_i' are truncated suitably so that their endpoints lie in S, as required in the statement of Lemma 2.5.10.

We now explain the algorithm that will yield the desired paths. We first test if γ_1' intersects some common peripheral disk with $\gamma_{m'}'$. If this is the case, then we consider the "first" such peripheral disk Q_{i_0}, and truncate the paths, so that the two resulting paths, denoted by γ_1 and γ_2, have an endpoint on ∂Q_{i_0}, but otherwise intersect disjoint sets of peripheral disks and γ_1 does not intersect Q_{i_0}. Then the statement of the lemma is proved with $m = 2$. The truncation is done in such a way that the left endpoint of γ_1' is the left endpoint of γ_1 and the right endpoint of $\gamma_{m'}'$ is the right endpoint of γ_2. Note here that subpaths of paths in $\mathcal{G}$ are also in $\mathcal{G}$.

If the above does not occur, then we test γ_1' against γ_{m-1}'. If they intersect some common peripheral disk, then we truncate them as above to obtain paths γ_1 and γ_2. Then we test γ_2 against γ_m' in the same way.

If the procedure does not stop we keep testing γ_1' against all paths, up to γ_2'. If the procedure still does not stop, we set $\gamma_1 = \gamma_1'$, and we start testing γ_2' against the other paths γ_i', $i = m, m - 1, \ldots, 3$, etc.

To finish the proof, one has to observe that the implemented truncation does not destroy the properties (i),(ii),(iv), and (v) in the statement of Lemma 2.5.10 with were true for the paths γ_i'. In particular, note that the peripheral disks in the second case of (v) have to be distinct by our algorithm.   □

Proof of Claim Suppose we are given closed paths $\tilde{\gamma}_1, \ldots, \tilde{\gamma}_m$ in the plane such that $\tilde{\gamma}_i$ starts at x_i and terminates at y_i, with $x_1 = x$ and $y_m = y$. The algorithm that we will use is very similar to the one used in the preceding proof of Lemma 2.5.10.

To illustrate the algorithm we assume that $m = 3$. We now check whether $\tilde{\gamma}_3$ intersects $\tilde{\gamma}_1$ or not. If $\tilde{\gamma}_3$ intersects $\tilde{\gamma}_1$, then we consider the first point $y_1' \in \tilde{\gamma}_3$ that $\tilde{\gamma}_1$ meets as it travels from $x = x_1$ to y_1. We call γ_1' the subpath of $\tilde{\gamma}_1$ from x_1 to y_1' (it could be that γ_1' is a constant path if $x_1 \in \tilde{\gamma}_3$). Then, we let γ_2' be the subpath of $\tilde{\gamma}_3$ from $y_2' := y_3 = y$ to $x_2' := y_1'$ (assuming that $\tilde{\gamma}_3$ is parametrized to travel from y_3 to x_3). The paths γ_1' and γ_2' share an endpoint but otherwise are disjoint, and they are the desired paths. Note that the alternative (b1) holds here.

If $\tilde{\gamma}_3$ does not intersect $\tilde{\gamma}_1$, then we check whether $\tilde{\gamma}_2$ intersects $\tilde{\gamma}_1$. If not, we set $\gamma_1' = \tilde{\gamma}_1$ and we also test if $\tilde{\gamma}_2$ intersects $\tilde{\gamma}_3$. Again, if this is not the case, then all three paths are disjoint, and thus we may set $\gamma_i' = \tilde{\gamma}_i$ for $i = 2, 3$ and the alternative (b2) holds. If $\tilde{\gamma}_2$ does intersect $\tilde{\gamma}_3$, we truncate $\tilde{\gamma}_2$ and $\tilde{\gamma}_3$ with the procedure we described in the previous paragraph, to obtain paths γ_2' and γ_3'. This procedure keeps the left endpoint of $\tilde{\gamma}_2$ as a left endpoint of γ_2' so the alternative (b2) holds between the paths γ_1' and γ_2'. For the paths γ_2' and γ_3' we have the alternative (b1).

Now, suppose that $\tilde{\gamma}_2$ intersects $\tilde{\gamma}_1$. We perform the truncating procedure as before, to obtain paths γ_1' and γ_2' that share an endpoint so the alternative (b1) holds here. The truncating procedure keeps the right endpoint of $\tilde{\gamma}_2$. Then we also need to test γ_2' against $\tilde{\gamma}_3$. If γ_2' does not intersect $\tilde{\gamma}_3$, then we set $\gamma_3' = \tilde{\gamma}_3$ and (b2) holds between γ_2' and γ_3'. Finally, if γ_2' intersects $\tilde{\gamma}_3$ then we truncate them and and the alternative (b1) holds.

This completes all possible cases, to obtain the desired paths that are disjoint, with the exceptions of some constant paths and of "consecutive" paths having a common endpoint.   □

Summarizing, in this section, we started with a discrete Sobolev function $\hat{f}$ and constructed a corresponding function $f : S \to \widehat{\mathbb{R}}$, defined on the entire carpet S, that also satisfies an upper gradient inequality.

2.5.2 Sobolev Spaces

Now, we proceed with the definition of the actual Sobolev spaces we will be using. Recall the Definition 2.5.8 of an upper gradient and also the definition of M_{Q_i} from Corollary 2.5.7.

Definition 2.5.11 Let $g : S \to \widehat{\mathbb{R}}$ be an extended function. We say that g lies in the *local weak* (resp. *strong*) *Sobolev space* $\mathcal{W}_{w,\mathrm{loc}}^{1,2}(S)$ (resp. $\mathcal{W}_{s,\mathrm{loc}}^{1,2}(S)$) if for every ball $B \subset\subset \Omega$ we have

$$\sum_{i \in I_B} M_{Q_i}(g)^2 \operatorname{diam}(Q_i)^2 < \infty, \qquad (2.18)$$

$$\sum_{i \in I_B} \underset{Q_i}{\operatorname{osc}}(g)^2 < \infty, \qquad (2.19)$$

and $\{\operatorname{osc}_{Q_i}(g)\}_{i \in \mathbb{N}}$ is a weak (resp. strong) upper gradient for g. Furthermore, if the above conditions hold for the full sums over $i \in \mathbb{N}$, then we say that f lies in the *weak* (resp. *strong*) *Sobolev space* $\mathcal{W}_w^{1,2}(S)$ (resp. $\mathcal{W}_s^{1,2}(S)$).

Again, in order to avoid making every time the distinction between weak and strong Sobolev spaces, we will also use the notation $\mathcal{W}_{*,\text{loc}}^{1,2}$, $\mathcal{W}_*^{1,2}$, where $*$ can mean either "w" or "s". See also the comments after Definition 2.5.1.

Note that part of the definition is that $|M_{Q_i}(g)| < \infty$ and $\operatorname{osc}_{Q_i}(g) < \infty$ for every peripheral disk Q_i, $i \in \mathbb{N}$. Also, each such function g comes with a family of good paths $\mathcal{G}_g$ defined as in the previous section, with the following properties: $\mathcal{H}^1(\gamma \cap S) = 0$, the upper gradient inequality for g holds along all subpaths of γ, and $\sum_{i:Q_i \cap \gamma' \neq \emptyset} \operatorname{osc}_{Q_i}(g) < \infty$ for subpaths γ' of γ compactly contained in Ω. In addition, in the case that $\sum_{i \in \mathbb{N}} \operatorname{osc}_{Q_i}(g)^2 < \infty$, we require that $\sum_{i:Q_i \cap \gamma \neq \emptyset} \operatorname{osc}_{Q_i}(\gamma) < \infty$ for all $\gamma \in \mathcal{G}_g$.

Remark 2.5.12 Observe that (2.18) is equivalent to

$$\sum_{i \in I_B} M_{Q_i}(|g|)^2 \operatorname{diam}(Q_i)^2 < \infty$$

if we assume (2.19). This is because $M_{Q_i}(|g|) \leq |M_{Q_i}(g)| + \operatorname{osc}_{Q_i}(g)$ and

$$\sum_{i \in I_B} \underset{Q_i}{\operatorname{osc}}(g)^2 \operatorname{diam}(Q_i)^2 < \infty.$$

Remark 2.5.13 The proof of Proposition 2.5.9 shows that if there exists a locally square-summable sequence $\{\rho(Q_i)\}_{i \in \mathbb{N}}$ that is an upper gradient for g, then $\{\operatorname{osc}_{Q_i}(g)\}_{i \in \mathbb{N}}$ is also an upper gradient for g.

Remark 2.5.14 As in the case of discrete Sobolev spaces, it also holds here that $\mathcal{W}_{s,\text{loc}}^{1,2}(S) \subset \mathcal{W}_{w,\text{loc}}^{1,2}(S)$; see (2.10).

Our discussion in Sect. 2.5.1 shows that each function $\hat{f} \in \widehat{\mathcal{W}}_{*,\text{loc}}^{1,2}(S)$ yields a function $f \in \mathcal{W}_{*,\text{loc}}^{1,2}(S)$ in a canonical way. Conversely, for any $g \in \mathcal{W}_{*,\text{loc}}^{1,2}(S)$ one can construct a discrete function $\hat{g}: \{Q_i\}_{i \in \mathbb{N}} \to \mathbb{R}$ by setting $\hat{g}(Q_i) = M_{Q_i}(g)$ for $i \in \mathbb{N}$. It is easy to check that $\hat{g}$ inherits the upper gradient inequality of g. Indeed, for a non-exceptional curve $\gamma \in \mathcal{G}_g$ that intersects Q_{i_1}, Q_{i_2}, let $x \in \partial Q_{i_1}$ be the point of last exit of γ from Q_{i_1}, and $y \in \partial Q_{i_2}$ be the point of first entry of γ in Q_{i_2}.

Then the subpath γ' of γ from x to y does not intersect Q_{i_1} or Q_{i_2}, so

$$|\hat{g}(Q_{i_1}) - \hat{g}(Q_{i_2})| \leq |M_{Q_{i_1}}(g) - g(x)| + |g(x) - g(y)| + |g(y) - M_{Q_{i_2}}(g)|$$

$$\leq \operatorname*{osc}_{Q_{i_1}}(g) + \sum_{i:Q_i \cap \gamma' \neq \emptyset} \operatorname*{osc}_{Q_i}(g) + \operatorname*{osc}_{Q_{i_2}}(g)$$

$$\leq \sum_{i:Q_i \cap \gamma \neq \emptyset} \operatorname*{osc}_{Q_i}(g). \tag{2.20}$$

Thus, $\hat{g} \in \widehat{\mathcal{W}}^{1,2}_{*,\mathrm{loc}}(S)$ with upper gradient $\{\operatorname*{osc}_{Q_i}(g)\}_{i \in \mathbb{N}}$, and there is a corresponding family of good paths $\mathcal{G}_{\hat{g}} \supset \mathcal{G}_g$.

Remark 2.5.15 Note that equality $\mathcal{G}_{\hat{g}} = \mathcal{G}_g$ does not hold in general, since changing the definition of g at one point $x \in S^{\circ}$ might destroy the upper gradient inequality of g for curves passing through that point. It turns out that these curves have weak and strong modulus zero. However, $\hat{g}$ is not affected by this change.

A question that arises is whether we can recover g from $\hat{g}$ using our previous construction in Sect. 2.5.1. As a matter of fact, this is the case, at least for points accessible by good paths $\gamma \in \mathcal{G}_g$. To prove this, let x be a point accessible by a good path γ, and Q_{i_0} be a peripheral disk with $Q_{i_0} \cap \gamma \neq \emptyset$; recall at this point Lemma 2.4.12 that allows us to approximate x by peripheral disks intersecting γ. Consider $y \in \partial Q_{i_0}$ to be the point of first entry of γ in Q_{i_0}, as γ travels from x to Q_{i_0}, and γ' be the subpath of γ from x to y. Then

$$|g(x) - \hat{g}(Q_{i_0})| = |g(x) - M_{Q_{i_0}}(g)| \leq |g(x) - g(y)| + |g(y) - M_{Q_{i_0}}(g)|$$

$$\leq \sum_{i:Q_i \cap \gamma' \neq \emptyset} \operatorname*{osc}_{Q_i}(g) + \operatorname*{osc}_{Q_{i_0}}(g).$$

As $Q_{i_0} \to x$ along γ, the right hand side converges to 0. This is because

$$\sum_{i:Q_i \cap \gamma \neq \emptyset} \operatorname*{osc}_{Q_i}(g) < \infty,$$

and the characteristic function of $\{i \in \mathbb{N} : Q_i \cap \gamma' \neq \emptyset\}$ converges to 0 as $Q_{i_0} \to x$. Thus,

$$g(x) = \liminf_{\substack{Q_i \to x \\ Q_i \cap \gamma \neq \emptyset}} \hat{g}(Q_i). \tag{2.21}$$

For non-accessible points of ∂Q_i we do not expect this, since g could have any value there between $m_{Q_i}(g)$ and $M_{Q_i}(g)$.

We now define a *normalized version* $\tilde{g}$ of g by using the construction in Sect. 2.5.1. More precisely, we define

$$\tilde{g}(x) = \liminf_{\substack{Q_i \to x \\ Q_i \cap \gamma \neq \emptyset}} \hat{g}(Q_i), \qquad (2.22)$$

whenever $x \in S$ is accessible by a path $\gamma \in \mathcal{G}_{\hat{g}}$, as in (2.11). For the other points of S we define $\tilde{g}$ as in the paragraph following (2.11). Note that the definition of $\tilde{g}$ depends on the good family $\mathcal{G}_{\hat{g}}$, which in turn depends on $\mathcal{G}_g$. By the discussion in Sect. 2.5.1, we have $\tilde{g} \in \mathcal{W}^{1,2}_{*,\mathrm{loc}}(S)$ with upper gradient $\{\mathrm{osc}_{Q_i}(\tilde{g})\}_{i\in\mathbb{N}}$, and the upper gradient inequality holds along paths of $\mathcal{G}_{\hat{g}} \supset \mathcal{G}_g$.

The function $\tilde{g}$ agrees with g for all points accessible by paths $\gamma \in \mathcal{G}_g$, as (2.21) shows. We remark that, by (2.20), $\{\mathrm{osc}_{Q_i}(g)\}_{i\in\mathbb{N}}$ is an upper gradient of $\hat{g}$, and by Corollary 2.5.7 we obtain that for the normalized version $\tilde{g}$ of g we always have

$$\mathrm{osc}_{Q_i}(\tilde{g}) \leq \mathrm{osc}_{Q_i}(g) \qquad (2.23)$$

for all $i \in \mathbb{N}$. The intuitive explanation is that the "jumps" at non-accessible points are precisely what makes a function non-normalized. Thus the process of normalization cuts these "jumps" and reduces the oscillation of the function. We summarize the above discussion in the following lemma:

Lemma 2.5.16 *For each function* $g \in \mathcal{W}^{1,2}_{*,\mathrm{loc}}(S)$ *there exists a function* $\tilde{g} \in \mathcal{W}^{1,2}_{*,\mathrm{loc}}(S)$ *such that*

  (i) $\tilde{g}$ *is defined by (2.22), where* $\hat{g}(Q_i) := M_{Q_i}(g)$ *for* $i \in \mathbb{N}$,
 (ii) $\tilde{g}$ *agrees with* g *on all points that are accessible by paths of* $\mathcal{G}_g$, *and*
(iii) $\mathrm{osc}_{Q_i}(\tilde{g}) \leq \mathrm{osc}_{Q_i}(g)$ *for all* $i \in \mathbb{N}$.

Moreover, the upper gradient inequality of $\tilde{g}$ *holds along paths of* $\mathcal{G}_{\hat{g}} \supset \mathcal{G}_g$.

The most important property of the normalized version $\tilde{g}$ of g is the following continuity property:

Lemma 2.5.17 *Let* $x \in S$ *be a point (not necessarily accessible). Then there exists a sequence of peripheral disks* Q_{i_n}, $n \in \mathbb{N}$, *converging to* x *(in the Hausdorff sense) such that* $M_{Q_{i_n}}(\tilde{g})$ *converges to* $\tilde{g}(x)$.

Proof It suffices to show that we can replace $\hat{g}(Q_i) = M_{Q_i}(g)$ in the definition (2.22) of $\tilde{g}$ by $M_{Q_i}(\tilde{g})$. Indeed, if x is an accessible point by $\gamma \in \mathcal{G}_{\hat{g}}$, then there exist arbitrarily small peripheral disks near x, intersecting γ, such that $\hat{g}(Q_i)$ approximates $\tilde{g}(x)$, by the definition of $\tilde{g}$. If x is non-accessible then $\tilde{g}(x)$ is defined through approximating x by accessible points, and hence we can find again small peripheral disks near x with the desired property.

Now we prove our claim. Let x be a point that is accessible by $\gamma \in G_{\hat{g}}$, so

$$\tilde{g}(x) = \liminf_{\substack{Q_i \to x \\ Q_i \cap \gamma \neq \emptyset}} M_{Q_i}(g).$$

We fix $i \in \mathbb{N}$. By an application of Lemma 2.4.3, we can find a point $y \in \partial Q_i$ that is accessible by a curve lying in G_g. By Lemma 2.5.16(ii) we have $\tilde{g}(y) = g(y)$. Hence, using Lemma 2.5.16(iii) we have

$$|M_{Q_i}(g) - M_{Q_i}(\tilde{g})| \leq |M_{Q_i}(g) - g(y)| + |\tilde{g}(y) - M_{Q_i}(\tilde{g})|$$

$$\leq \operatorname*{osc}_{Q_i}(g) + \operatorname*{osc}_{Q_i}(\tilde{g}) \leq 2 \operatorname*{osc}_{Q_i}(g).$$

Since $\{\operatorname{osc}_{Q_i}(g)\}_{i \in \mathbb{N}}$ is locally square-summable near x, and the peripheral disks Q_i become arbitrarily small as $Q_i \to x$ and $Q_i \cap \gamma \neq \emptyset$, it follows that

$$\limsup_{\substack{Q_i \to x \\ Q_i \cap \gamma \neq \emptyset}} |M_{Q_i}(g) - M_{Q_i}(\tilde{g})| \leq \limsup_{\substack{Q_i \to x \\ Q_i \cap \gamma \neq \emptyset}} 2 \operatorname*{osc}_{Q_i}(g) = 0.$$

This shows that we can indeed define

$$\tilde{g}(x) = \liminf_{\substack{Q_i \to x \\ Q_i \cap \gamma \neq \emptyset}} M_{Q_i}(\tilde{g})$$

as desired.   □

This discussion suggests that we identify the functions g of the space $\mathcal{W}^{1,2}_{*,\mathrm{loc}}(S)$ that have the "same" normalized version $\tilde{g}$. This will be made more precise with the following lemma:

Lemma 2.5.18 *Let $f, g \in \mathcal{W}^{1,2}_{*,\mathrm{loc}}(S)$, each of them coming with a family of good paths G_f and G_g, respectively. The following are equivalent:*

(a) *There exists a family G_0 that contains almost every path in Ω (i.e., the complement of G_0 has carpet modulus zero) such that $f(x) = g(x)$ for all points x that are accessible by paths $\gamma \in G_0$.*
(b) *For the normalized versions $\tilde{f}$ and $\tilde{g}$ and for all $i \in \mathbb{N}$ we have*

$$M_{Q_i}(\tilde{f}) = M_{Q_i}(\tilde{g}), \quad m_{Q_i}(\tilde{f}) = m_{Q_i}(\tilde{g}), \quad \text{and} \quad \operatorname*{osc}_{Q_i}(\tilde{f}) = \operatorname*{osc}_{Q_i}(\tilde{g}).$$

Recall that the carpet modulus in (a) is either weak or strong, depending on the Sobolev space that f, g lie in.

Proof First, we assume that there exists a family G_0 such that $f(x) = g(x)$ for all points x accessible by $\gamma \in G_0$. Then for points x accessible by paths in $G := G_f \cap G_g \cap G_0$ we have $\tilde{f}(x) = f(x) = g(x) = \tilde{g}(x)$ by Lemma 2.5.16(ii). Note that

$\mathcal{G}$ contains almost every path, by the subadditivity of modulus. Fix a peripheral disk Q_{i_0}, $\varepsilon > 0$, and a point $x \in \partial Q_{i_0}$ with $|M_{Q_{i_0}}(\tilde{f}) - \tilde{f}(x)| < \varepsilon$; recall the definition of $M_{Q_{i_0}}(\tilde{f})$. Then, consider a point $x' \in \partial Q_{i_0}$ near x that is accessible by a curve $\gamma \in \mathcal{G}_{\hat{f}}$, such that $|\tilde{f}(x) - \tilde{f}(x')| < \varepsilon$; this can be done by the definition of $\tilde{f}$ on non-accessible points. Now, Lemma 2.4.7(a) yields a circular arc γ_r around x' with a small $r > 0$ (see also Fig. 2.2) such that:

(1) $\gamma_r \cap Q_{i_0} = \emptyset$ and γ_r has its endpoints on ∂Q_{i_0}, so that γ_r defines a crosscut separating x' from ∞ in $\mathbb{R}^2 \setminus Q_{i_0}$,
(2) γ_r avoids the set $\gamma \cap S$ and intersects a peripheral disk Q_{i_1}, $i_1 \neq i_0$, that is also intersected by γ,
(3) $\gamma_r \in \mathcal{G}$,
(4) $\sum_{i : Q_i \cap \gamma_r \neq \emptyset} \operatorname{osc}_{Q_i}(\tilde{f}) < \varepsilon$.

Let $y \in \partial Q_{i_0} \cap \gamma_r$ a point accessible by γ_r, i.e., an endpoint of γ_r. Then $\tilde{f}(y) = \tilde{g}(y)$, because $\gamma_r \in \mathcal{G}$. Let γ' be the subpath of γ from Q_{i_1} to x. Using the upper gradient inequality of $\tilde{f}$, which holds along $\gamma' \in \mathcal{G}_{\hat{f}}$ and $\gamma_r \in \mathcal{G} \subset \mathcal{G}_f \subset \mathcal{G}_{\hat{f}}$, we have

$$|\tilde{f}(x') - \tilde{f}(y)| \leq \sum_{i : Q_i \cap \gamma' \neq \emptyset} \operatorname{osc}_{Q_i}(\tilde{f}) + \sum_{i : Q_i \cap \gamma_r \neq \emptyset} \operatorname{osc}_{Q_i}(\tilde{f}) \leq \sum_{i : Q_i \cap \gamma' \neq \emptyset} \operatorname{osc}_{Q_i}(\tilde{f}) + \varepsilon.$$

If Q_{i_1} is sufficiently close to x (and thus $r > 0$ is chosen to be smaller), then the last sum can be made less than ε. Putting all the estimates together we obtain

$$|M_{Q_{i_0}}(\tilde{f}) - \tilde{g}(y)| < 4\varepsilon.$$

This shows that $M_{Q_{i_0}}(\tilde{f}) \leq M_{Q_{i_0}}(\tilde{g})$. Interchanging the roles of $\tilde{f}$ and $\tilde{g}$ we obtain the equality. The equality $m_{Q_{i_0}}(\tilde{f}) = m_{Q_{i_0}}(\tilde{g})$ is proved by using the same argument.

For the converse, note that for all points x that are accessible by paths $\gamma \in \mathcal{G}_{\hat{g}}$ we have

$$\tilde{g}(x) = \liminf_{\substack{Q_i \to x \\ Q_i \cap \gamma \neq \emptyset}} M_{Q_i}(\tilde{g}),$$

and an analogous statement is true for $\tilde{f}$; see Lemma 2.5.17 and its proof. Hence, for all points x accessible by paths $\gamma \in \mathcal{G}_0 := \mathcal{G}_f \cap \mathcal{G}_g \subset \mathcal{G}_{\hat{f}} \cap \mathcal{G}_{\hat{g}}$ we have

$$\tilde{g}(x) = \liminf_{\substack{Q_i \to x \\ Q_i \cap \gamma \neq \emptyset}} M_{Q_i}(\tilde{g}) = \liminf_{\substack{Q_i \to x \\ Q_i \cap \gamma \neq \emptyset}} M_{Q_i}(\tilde{f}) = \tilde{f}(x).$$

On the other hand, for such points we also have $\tilde{f}(x) = f(x)$ and $\tilde{g}(x) = g(x)$ by Lemma 2.5.16(ii). The conclusion follows, if one notes that the curve family $\mathcal{G}_0$ contains almost every path in Ω, by the subadditivity of modulus. $\qquad\square$

Hence, we can identify functions $f, g \in \mathcal{W}^{1,2}_{*,\mathrm{loc}}(S)$ whenever their normalized versions $\tilde{f}, \tilde{g}$ yield the same sequences $\{M_{Q_i}(\tilde{f})\}_{i\in\mathbb{N}}$, $\{\mathrm{osc}_{Q_i}(\tilde{f})\}_{i\in\mathbb{N}}$. The identification allows us to regard $\mathcal{W}^{1,2}_{w,\mathrm{loc}}(S)$ and $\mathcal{W}^{1,2}_{s,\mathrm{loc}}(S)$ as subsets of a space of sequences, the (non-linear) correspondence being

$$g \mapsto (\{M_{Q_i}(\tilde{g})\,\mathrm{diam}(Q_i)\}_{i\in\mathbb{N}}, \{\underset{Q_i}{\mathrm{osc}}(\tilde{g})\}_{i\in\mathbb{N}}).$$

These sequences are locally square-summable in the sense of (2.18) and (2.19). If g were originally in one of the non-local Sobolev spaces instead, then we could identify g with an element of $\ell^2(\mathbb{N}) \times \ell^2(\mathbb{N})$.

However, we will not use this identification in the next sections, and the Sobolev functions that we use are not necessarily normalized, unless otherwise stated.

2.5.3 Examples

We give some examples of functions lying in the Sobolev spaces $\mathcal{W}^{1,2}_{w,\mathrm{loc}}(S)$ and $\mathcal{W}^{1,2}_{s,\mathrm{loc}}(S)$.

Example 2.5.19 Let $f: \Omega \to \mathbb{R}$ be a locally Lipschitz function, i.e., for every compact set $K \subset \Omega$ there exists a constant $L > 0$ such that $|f(x)-f(y)| \leq L|x-y|$ for $x, y \in K$. Then $f|_S \in \mathcal{W}^{1,2}_{s,\mathrm{loc}}(S) \subset \mathcal{W}^{1,2}_{w,\mathrm{loc}}(S)$. In particular, this is true if $f: \Omega \to \mathbb{R}$ is smooth.

To see this, consider a compact exhaustion $\{K_n\}_{n\in\mathbb{N}}$, $K_n \subset K_{n+1}$ of Ω, and an increasing sequence of Lipschitz constants L_n for $f|_{K_n}$. Define $\rho(Q_i) = L_n \mathrm{diam}(Q_i)$ where $n \in \mathbb{N}$ is the smallest integer such that $Q_i \subset K_n$. If $B \subset\subset \Omega$, then f is bounded on B, thus

$$\sum_{i\in I_B} M_{Q_i}(f)^2 \,\mathrm{diam}(Q_i)^2 < \infty,$$

by Corollary 2.3.6. Also, $\mathrm{osc}_{Q_i}(f) \leq \rho(Q_i)$ by the Lipschitz condition, and if $B \subset K_N$, then

$$\sum_{i\in I_B} \rho(Q_i)^2 \leq L_N^2 \sum_{i\in I_B} \mathrm{diam}(Q_i)^2 < \infty.$$

Finally, let γ be a curve in Ω with $\mathcal{H}^1(\gamma \cap S) = 0$, and $x, y \in \gamma \cap S$. We wish to show that

$$|f(x) - f(y)| \leq \sum_{i:Q_i \cap \gamma \neq \emptyset} \rho(Q_i).$$

If suffices to prove this for a closed subpath of γ that connects x and y, which we still denote by γ. This will imply that $\{\rho(Q_i)\}_{i \in \mathbb{N}}$ is a strong upper gradient for f, and thus, so is $\{\mathrm{osc}_{Q_i}(f)\}_{i \in \mathbb{N}}$; see Remark 2.5.13.

For fixed $\varepsilon > 0$ we cover the compact set $\gamma \cap S$ with finitely many balls B_j of radius r_j such that $\sum_j r_j < \varepsilon$. Furthermore, we may assume that $B_j \subset\subset \Omega$ for all j. Then there are at most finitely many peripheral disks that intersect γ and are not covered by $\bigcup_j B_j$. Indeed, note that the closure of each of these peripheral disks must intersect both $\partial(\bigcup_j B_j)$ and $\gamma \cap S$. On the other hand, $\partial(\bigcup_j B_j)$ and $\gamma \cap S$ have positive distance, hence the peripheral disks whose closure intersects both of them have diameters bounded below. By Lemma 2.3.4 we conclude that these peripheral disks have to be finitely many.

We let A_k, $k \in \{1, \ldots, M\}$, be the joint collection of the balls B_j and of the finitely many peripheral disks Q_i not covered by $\bigcup_j B_j$. In other words, for each $k = 1, \ldots, M$ we have $A_k = B_j$ for some j, or A_k is one of these peripheral disks, and the sets A_k are distinct, i.e., we do not include a ball or a peripheral disk twice. Note that x and y lie in some balls, so after reordering we suppose that $x \in A_1$ and $y \in A_M$. The union $\bigcup_k A_k$ contains the curve γ, and thus it contains a connected chain of sets A_k, connecting x to y. We consider a minimal sub-collection of $\{A_k\}_k$ that connects x and y (there are only finitely many sub-collections), and we still denote it in the same way. Then, using the minimality, we can order the sets A_k in such a way, that $x \in A_1$, $y \in A_M$, and $A_k \cap A_l \neq \emptyset$ if and only if $l = k \pm 1$.

Now, we choose points $x_{k+1} \in A_k \cap A_{k+1}$ for $k = 1, \ldots, M - 1$, and set $x_1 = x \in A_1$ and $x_{M+1} = y \in A_M$. Suppose that $\bigcup_k A_k \subset K_N$ for some $N \in \mathbb{N}$. If A_k is a ball B_j, then $|f(x_k) - f(x_{k+1})| \leq 2L_N r_j$, and if A_k is a peripheral disk Q_i, then $|f(x_k) - f(x_{k+1})| \leq \rho(Q_i)$; see definition of $\rho(Q_i)$. Hence,

$$|f(x) - f(y)| \leq \sum_{k=1}^{M} |f(x_k) - f(x_{k+1})|$$

$$\leq 2L_N \sum_j r_j + \sum_{i:Q_i \cap \gamma \neq \emptyset} \rho(Q_i)$$

$$\leq 2L_N \varepsilon + \sum_{i:Q_i \cap \gamma \neq \emptyset} \rho(Q_i).$$

Letting $\varepsilon \to 0$ shows that $\{\rho(Q_i)\}_{i \in \mathbb{N}}$ is a strong upper gradient for f, as desired.

Example 2.5.20 Let $f: \Omega \to f(\Omega) \subset \mathbb{C}$ be a homeomorphism that lies in the classical space $W_{\mathrm{loc}}^{1,2}(\Omega)$. Then $f|_S \in \mathcal{W}_{s,\mathrm{loc}}^{1,2}(S)$ in the sense that this holds for the real and imaginary parts of $f|_S$. In particular, (locally) quasiconformal maps on Ω lie in $\mathcal{W}^{1,2}_{s,\text{loc}}(S)$; see Sect. 3.3.1 in Chap. 3, and also [2] for the definition of a quasiconformal map and basic properties.

Since f is locally bounded, for any ball $B \subset\subset \Omega$ we have

$$\sum_{i \in I_B} M_{Q_i}(|f|)^2 \operatorname{diam}(Q_i)^2 < \infty.$$

We will show that $\rho(Q_i) := \operatorname{diam}(f(Q_i))$ is also locally square-summable, and it is a strong upper gradient for f, i.e., there exists a path family Γ_0 in Ω with $\operatorname{mod}_s(\Gamma_0) = 0$ such that

$$|f(x) - f(y)| \leq \sum_{i:Q_i \cap \gamma \neq \emptyset} \rho(Q_i)$$

for $\gamma \subset \Omega$, $\gamma \notin \Gamma_0$, and $x, y \in \gamma \cap S$. Note that this suffices by Remark 2.5.13, and it will imply that $\{\operatorname{osc}_{Q_i}(\operatorname{Re}(f))\}_{i \in \mathbb{N}}$ and $\{\operatorname{osc}_{Q_i}(\operatorname{Im}(f))\}_{i \in \mathbb{N}}$ are strong upper gradients for $\operatorname{Re}(f)$ and $\operatorname{Im}(f)$, respectively. Using a compact exhaustion, we see that it suffices to show this for paths γ contained in an open set $V \subset\subset \Omega$.

Let U be a neighborhood of V such that $V \subset\subset U \subset\subset \Omega$, and U contains all peripheral disks that intersect V. By a recent theorem of Iwaniec, Kovalev and Onninen [23, Theorem 1.2] there exist smooth homeomorphisms $f_n : U \to f_n(U)$ such that $f_n \to f$ uniformly on U and in $W^{1,2}(U)$. For each $n \in \mathbb{N}$ the function f_n is locally Lipschitz, so by Example 2.5.19 it satisfies

$$|f_n(x) - f_n(y)| \leq \sum_{i:Q_i \cap \gamma \neq \emptyset} \operatorname*{osc}_{Q_i}(f_n) = \sum_{i:Q_i \cap \gamma \neq \emptyset} \operatorname{diam}(f_n(Q_i)) \qquad (2.24)$$

for all $\gamma \subset V$ with $\mathcal{H}^1(\gamma \cap S) = 0$ and $x, y \in \gamma \cap S$. We claim that the sequence $\{\operatorname{diam}(f_n(Q_i))\}_{i \in I_V}$ converges in ℓ^2. Suppose for the moment that this is the case. Then the ℓ^2 limit of that sequence must be the same as the pointwise limit, namely $\{\operatorname{diam}(f(Q_i))\}_{i \in I_V}$, so the latter is also square-summable. We pass to a subsequence as required in Fuglede's Lemma 2.3.8 and then we take limits in (2.24), so as to obtain

$$|f(x) - f(y)| \leq \sum_{i:Q_i \cap \gamma \neq \emptyset} \operatorname{diam}(f(Q_i))$$

for all paths $\gamma \subset V$ outside an exceptional family Γ_0 with $\operatorname{mod}_s(\Gamma_0) = 0$, and $x, y \in \gamma \cap S$.

Now, we show our claim. By the quasiroundness assumption, there exist balls $B(x_i, r_i) \subset Q_i \subset B(x_i, R_i)$, $i \in \mathbb{N}$, with $R_i/r_i \leq K_0$. Since $\operatorname{dist}(V, \partial U) > 0$, there exist only finitely many peripheral disks Q_i, $i \in I_V$, such that $B(x_i, 2R_i)$ is not contained in U; see Lemma 2.3.4. Let J be the family of such indices. Since this is a finite set and $\mathrm{diam}(f_n(Q_i)) \to \mathrm{diam}(f(Q_i))$ all $i \in J$, it suffices to show that $\{\mathrm{diam}(f_n(Q_i))\}_{i \in I_V \setminus J}$ converges to $\{\mathrm{diam}(f(Q_i))\}_{i \in I_V \setminus J}$ in ℓ^2.

We fix $i \in I_V \setminus J$. For each $r \in [R_i, 2R_i]$ and $x, y \in \partial Q_i \subset B(x_i, r)$, by the maximum principle applied to the homeomorphisms f_n and the fundamental theorem of calculus we have

$$|f_n(x) - f_n(y)| \leq \underset{\partial B(x_i, r)}{\mathrm{osc}} (f_n) \leq \int_{\partial B(x_i, r)} |\nabla f_n| \, ds,$$

where $\nabla f_n = (\nabla \mathrm{Re}(f_n), \nabla \mathrm{Im}(f_n))$. Integrating over $r \in [R_i, 2R_i]$ we obtain

$$|f_n(x) - f_n(y)| \leq C R_i \fint_{B(x_i, 2R_i)} |\nabla f_n|$$

for some constant $C > 0$. Since $x, y \in \partial Q_i$ were arbitrary, we have

$$\mathrm{diam}(f_n(Q_i)) \leq C R_i \fint_{B(x_i, 2R_i)} |\nabla f_n| \tag{2.25}$$

for all $i \in I_V \setminus J$.

In the following, we restrict the domain of f to U, and by assumption we have $f_n \to f$ in $W^{1,2}(U)$. Using the uncentered maximal function $M(g)(x) = \sup_{x \in B} \fint_B |g|$ to change the center of balls we have

$$\left| R_i \fint_{B(x_i, 2R_i)} |\nabla f_n| - R_i \fint_{B(x_i, 2R_i)} |\nabla f| \right| \leq R_i \fint_{B(x_i, 2R_i)} |\nabla f_n - \nabla f|$$

$$\leq C R_i \fint_{B(x_i, r_i)} M(|\nabla f_n - \nabla f|),$$

where the constant $C > 0$ depends only on the quasiroundness constant K_0. Thus, for some constants $C', C'', C''' > 0$ we have

$$\sum_{i \in I_V \setminus J} \left| R_i \fint_{B(x_i, 2R_i)} |\nabla f_n| - R_i \fint_{B(x_i, 2R_i)} |\nabla f| \right|^2$$

$$\leq C' \sum_{i \in I_V \setminus J} R_i^2 \left(\fint_{B(x_i, r_i)} M(|\nabla f_n - \nabla f|) \right)^2$$

$$\leq C' \sum_{i \in I_V \setminus J} R_i^2 \fint_{B(x_i, r_i)} M(|\nabla f_n - \nabla f|)^2$$

$$\leq C'' \sum_{i \in I_V \setminus J} \int_{B(x_i, r_i)} M(|\nabla f_n - \nabla f|)^2$$

$$\leq C'' \int_U M(|\nabla f_n - \nabla f|)^2$$

$$\leq C''' \int_U |\nabla f_n - \nabla f|^2.$$

This shows that $\{R_i f_{B(x_i,2R_i)} |\nabla f_n|\}_{i \in I_V \setminus J}$ converges to $\{R_i f_{2B(x_i,R_i)} |\nabla f|\}_{i \in I_V \setminus J}$ in ℓ^2. Since this sequence dominates $\{\mathrm{diam}(f_n(Q_i))\}_{i \in I_V \setminus J}$ by (2.25), it follows that $\{\mathrm{diam}(f_n(Q_i))\}_{i \in I_V \setminus J}$ converges in ℓ^2 as well, as claimed.

Remark 2.5.21 One could do the preceding proof directly for quasiconformal maps, by approximating them with smooth quasiconformal maps; this is a more elementary result than the approximation of Sobolev homeomorphisms by smooth homeomorphisms. However, the use of the strong result [23, Theorem 1.2] proves a more general result while the proof remains essentially the same. Moreover, more recently the current author has established the approximation of continuous *monotone* $W^{1,2}$ functions by smooth monotone $W^{1,2}$ functions [30]. Here, a function is monotone if it satisfies the maximum and minimum principles. Using the approximation result of [30], the above proof can be applied with no changes to show that continuous monotone $W^{1,2}$ functions on Ω lie in $\mathcal{W}_s^{1,2}(S)$.

Remark 2.5.22 The same proof shows that if $\overline{\Omega}$ is compact and f is quasiconformal in a neighborhood of $\overline{\Omega}$ then $f|_S \in \mathcal{W}_s^{1,2}(S)$.

Remark 2.5.23 From the above proof we see that if $f_n \in \mathcal{W}_{s,\mathrm{loc}}^{1,2}(S)$ converges to $f : S \to \mathbb{R}$ locally uniformly and for each $V \subset\subset \Omega$ the sequence $\{\mathrm{osc}_{Q_i}(f_n)\}_{i \in I_V}$ converges to $\{\mathrm{osc}_{Q_i}(f)\}_{i \in I_V}$ in ℓ^2, then f lies in $\mathcal{W}_{s,\mathrm{loc}}^{1,2}(S)$.

2.5.4 Pullback of Sobolev Spaces

Here we study the invariance of Sobolev spaces under quasiconformal maps between relative Sierpiński carpets.

Let (S, Ω), (S', Ω') be two relative Sierpiński carpets and let $F : \Omega' \to \Omega$ be a locally quasiconformal homeomorphism that maps the peripheral disks Q'_i of S' to the peripheral disks $Q_i = F(Q'_i)$ of S.

Proposition 2.5.24 *If $g \in \mathcal{W}_{w,\mathrm{loc}}^{1,2}(S)$, then the pullback $g \circ F$ lies in $\mathcal{W}_{w,\mathrm{loc}}^{1,2}(S')$.*

Proof Let $g \in \mathcal{W}_{w,\mathrm{loc}}^{1,2}(S)$, and note that $M_{Q'_i}(g \circ F) = M_{Q_i}(g)$ and $\mathrm{osc}_{Q'_i}(g \circ F) = \mathrm{osc}_{Q_i}(g)$. We only have to show that there exists a path family Γ'_0 in Ω' with weak modulus equal to zero such that

$$|g \circ F(x) - g \circ F(y)| \leq \sum_{i : Q'_i \cap \gamma \neq \emptyset} \mathrm{osc}_{Q'_i}(g \circ F) = \sum_{i : Q_i \cap F(\gamma) \neq \emptyset} \mathrm{osc}_{Q_i}(g) \qquad (2.26)$$

whenever $\gamma \subset \Omega'$, $\gamma \notin \Gamma'_0$, and $x, y \in \gamma \cap S$.

By our assumption on g, there exists a path family Γ_0 in Ω with $\mathrm{mod}_w(\Gamma_0) = 0$ such that the upper gradient inequality for g holds along paths $\gamma \notin \Gamma_0$. By the equality in (2.26) it suffices to show that for almost every γ in Ω' the image $F(\gamma)$ avoids the exceptional family Γ_0. Equivalently, we show that the family $\Gamma_0' := F^{-1}(\Gamma_0)$ has weak modulus equal to zero.

Note that $\mathrm{mod}_w(\Gamma_0) = 0$ implies that $\mathrm{mod}_2(\Gamma_0) = 0$ by Lemma 2.3.3. Since F is locally quasiconformal, so is F^{-1}, and they preserve conformal modulus zero. Therefore, $\mathrm{mod}_2(F^{-1}(\Gamma_0)) = 0$. Again, by Lemma 2.3.3 we have $\mathrm{mod}_w(\Gamma_0') = 0$, as desired. $\square$

Corollary 2.5.25 *Assume that the peripheral circles $\partial Q_i'$, ∂Q_i of S', S, respectively, are uniform quasicircles and let $F: S' \to S$ be a local quasisymmetry. Then for any $g \in \mathcal{W}_{w,\mathrm{loc}}^{1,2}(S)$ we have $g \circ F \in \mathcal{W}_{w,\mathrm{loc}}^{1,2}(S')$.*

See [21, Chaps. 10–11] for background on quasisymmetric maps. The corollary follows immediately from Proposition 2.5.24 and the following lemma:

Lemma 2.5.26 *Assume that the peripheral circles $\partial Q_i'$, ∂Q_i of S', S, respectively, are uniform quasicircles and let $F: S' \to S$ be a local quasisymmetry. Then F extends to a locally quasiconformal map from Ω' onto Ω.*

We only provide a sketch of the proof.

Proof of Lemma 2.5.26 The first observation is that since F is a homeomorphism it maps each peripheral circle $\partial Q_i'$ of S' onto a peripheral circle ∂Q_i of S; see [7, Lemma 5.5] for an argument. Then one uses the well-known Beurling–Ahlfors extension to obtain a quasiconformal extension $F: Q_i' \to Q_i$ inside each peripheral disk. The resulting map $F: \Omega' \to \Omega$ will be locally quasiconformal and the proof can be found in [7, Sect. 5], where careful quantitative estimates are also shown. In our case we do not need such careful estimates. $\square$

2.5.5 Properties of Sobolev Spaces

We record here some properties of Sobolev functions:

Proposition 2.5.27 *The space $\mathcal{W}_{*,\mathrm{loc}}^{1,2}(S)$ is linear. Moreover, if $u, v \in \mathcal{W}_{*,\mathrm{loc}}^{1,2}(S)$, then the following functions also lie in $\mathcal{W}_{*,\mathrm{loc}}^{1,2}(S)$:*

(a) $|u|$, with $\mathrm{osc}_{Q_i}(|u|) \leq \mathrm{osc}_{Q_i}(u)$,
(b) $u \vee v := \max(u, v)$, with $\mathrm{osc}_{Q_i}(u \vee v) \leq \max\{\mathrm{osc}_{Q_i}(u), \mathrm{osc}_{Q_i}(v)\}$,
(c) $u \wedge v := \min(u, v)$, with $\mathrm{osc}_{Q_i}(u \wedge v) \leq \max\{\mathrm{osc}_{Q_i}(u), \mathrm{osc}_{Q_i}(v)\}$,

where $i \in \mathbb{N}$. Moreover,

(d) *if u and v are locally bounded in S, then $u \cdot v$ lies in the corresponding Sobolev space, with*

$$\underset{Q_i}{\operatorname{osc}}(u \cdot v) \leq M_{Q_i}(|v|) \underset{Q_i}{\operatorname{osc}}(u) + M_{Q_i}(|u|) \underset{Q_i}{\operatorname{osc}}(v).$$

for all $i \in \mathbb{N}$.

Finally, if we set $f = u \vee v$ and $g = u \wedge v$, then we have the inequality

$$\underset{Q_i}{\operatorname{osc}}(f)^2 + \underset{Q_i}{\operatorname{osc}}(g)^2 \leq \underset{Q_i}{\operatorname{osc}}(u)^2 + \underset{Q_i}{\operatorname{osc}}(v)^2 \tag{2.27}$$

for all $i \in \mathbb{N}$.

Proof To prove that the spaces are linear, we note that if u, v are Sobolev functions and $a, b \in \mathbb{R}$, then $\operatorname{osc}_{Q_i}(au + bv) \leq |a| \operatorname{osc}_{Q_i}(u) + |b| \operatorname{osc}_{Q_i}(v)$, which shows that the upper gradient inequalities of u and v yield an upper gradient inequality for $au + bv$. The summability conditions (2.18) and (2.19) in the definition of a Sobolev function are trivial.

Part (a) follows from the triangle inequality $||u(x)| - |u(y)|| \leq |u(x) - u(y)|$, which shows that $|u|$ inherits its upper gradient inequality from u.

To show (b) note that $u \vee v = (u + v + |u - v|)/2$. Using the linearity of Sobolev spaces and part (a) we obtain that $u \vee v$ also lies in the Sobolev space. To show the inequality, we only need to observe that

$$M_{Q_i}(u \vee v) = \max\{M_{Q_i}(u), M_{Q_i}(v)\} \quad \text{and}$$
$$m_{Q_i}(u \vee v) \geq \max\{m_{Q_i}(u), m_{Q_i}(v)\}. \tag{2.28}$$

Part (c) is proved in the exact same way, if one notes that

$$M_{Q_i}(u \wedge v) \leq \min\{M_{Q_i}(u), M_{Q_i}(v)\} \quad \text{and}$$
$$m_{Q_i}(u \wedge v) = \min\{m_{Q_i}(u), m_{Q_i}(v)\}. \tag{2.29}$$

For part (d) we note that the oscillation inequality is a straightforward computation. This, together with the local boundedness of u and v show immediately the summability conditions (2.18) and (2.19); see also Corollary 2.3.6. We only have to show the upper gradient inequality. Suppose that $\gamma \subset\subset \Omega$ is a path that is good for both u and v, and connects $x, y \in S$. Since $\gamma \subset\subset \Omega$, there exists $M > 0$ such that $|u| \leq M, |v| \leq M$ on $\gamma \cap S$. Then

$$|u(x)v(x) - u(y)v(y)| \leq M|u(x) - u(y)| + M|v(x) - v(y)|$$

$$\leq M \sum_{i: Q_i \cap \gamma \neq \emptyset} \left(\underset{Q_i}{\operatorname{osc}}(u) + \underset{Q_i}{\operatorname{osc}}(v)\right).$$

Since this also holds for subpaths of γ, the proof of Proposition 2.5.9 shows that

$$|u(x)v(x) - u(y)v(y)| \leq \sum_{i:Q_i \cap \gamma \neq \emptyset} \underset{Q_i}{\mathrm{osc}}(u \cdot v),$$

as desired; see also Remark 2.5.13.

To show inequality (2.27) we fix $i \in \mathbb{N}$, and for simplicity drop Q_i from the notations $\mathrm{osc}_{Q_i}, M_{Q_i}, m_{Q_i}$. We now split into cases, and by symmetry we only have to check two cases. If $M(u) \geq M(v)$ and $m(u) \geq m(v)$, then by (2.28) and (2.29) we have

$$\mathrm{osc}(f)^2 + \mathrm{osc}(g)^2 \leq (M(u) - m(u))^2 + (M(v) - m(v))^2 = \mathrm{osc}(u)^2 + \mathrm{osc}(v)^2.$$

If $M(u) \geq M(v)$ and $m(u) \leq m(v)$, then using again (2.28) and (2.29) we obtain

$$\mathrm{osc}(f)^2 + \mathrm{osc}(g)^2 \leq (M(u) - m(v))^2 + (M(v) - m(u))^2$$
$$= \mathrm{osc}(u)^2 + \mathrm{osc}(v)^2 - 2(M(u) - M(v))(m(v) - m(u)).$$

The last term in the above expression is non-negative by assumption, thus the expression is bounded by $\mathrm{osc}(u)^2 + \mathrm{osc}(v)^2$, as claimed.   □

Next, we include a lemma that allows us to "patch" together Sobolev functions. For an open set $V \subset \Omega$ recall that $\partial_* V = \partial V \cap S$; see Fig. 2.3. This notation is not to be confused with $\mathcal{W}_*^{1,2}$, where $*$ stands for "w" or "s".

Lemma 2.5.28 *Let $V \subset \Omega$ be an open set such that $\partial_* V \neq \emptyset$. Let $\phi, \psi \in \mathcal{W}_{*,\mathrm{loc}}^{1,2}(S)$ such that $\phi = \psi$ on $\partial_* V$. Then $h := \phi \chi_{S \cap V} + \psi \chi_{S \setminus V} \in \mathcal{W}_{*,\mathrm{loc}}^{1,2}(S)$.*
   Moreover, for each $i \in \mathbb{N}$ we have $\mathrm{osc}_{Q_i}(h) = \mathrm{osc}_{Q_i}(\phi)$ if $\partial Q_i \subset V$, $\mathrm{osc}_{Q_i}(h) = \mathrm{osc}_{Q_i}(\psi)$ if $\partial Q_i \subset \Omega \setminus \overline{V}$, and $\mathrm{osc}_{Q_i}(h) \leq \mathrm{osc}_{Q_i}(\phi) + \mathrm{osc}_{Q_i}(\psi)$ otherwise.

Proof We first show the oscillation relations. The first two are trivial and for the last one we note that if $\partial Q_i \not\subset V$ and $\partial Q_i \not\subset \Omega \setminus \overline{V}$, then there exists a point $x \in \partial Q_i \cap \partial V \subset \partial_* V$, so $\phi(x) = \psi(x)$. Let $z, w \in \partial Q_i$ be arbitrary. If $h(z) = \phi(z)$

Fig. 2.3 An open set V (pink) intersecting a round carpet S, and the set $\partial_* V$ that corresponds to V. Here, Ω has one boundary component, the largest circle. Also, $\overline{S}$ is an actual Sierpiński carpet, as defined in the introduction, Chap. 1

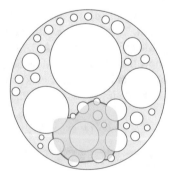

and $h(w) = \psi(w)$ then

$$|h(z) - h(w)| \le |h(z) - h(x)| + |h(x) - h(w)|$$
$$= |\phi(z) - \phi(x)| + |\psi(x) - \psi(w)|$$
$$\le \operatorname*{osc}_{Q_i}(\phi) + \operatorname*{osc}_{Q_i}(\psi).$$

The above inequality also holds trivially in the case $h(z) = \phi(z)$ and $h(w) = \phi(w)$, or $h(z) = \psi(z)$ and $h(w) = \psi(w)$. This proves the claim.

The summability condition (2.9) follows immediately from the oscillation relations, and the summability condition (2.8) follows from the fact that $M_{Q_i}(|h|) \le M_{Q_i}(|\phi|) + M_{Q_i}(|\psi|)$; see Remark 2.5.12. It remains to show the upper gradient inequality. By Remark 2.5.13, it suffices to prove that there exists a locally square-summable sequence $\{\rho(Q_i)\}_{i \in \mathbb{N}}$ that is an upper gradient for h. This will imply that $\{\operatorname{osc}_{Q_i}(h)\}_{i \in \mathbb{N}}$ has the same property.

Let γ be a path that is good for both ϕ and ψ and joins $x, y \in \gamma \cap S$. We wish to prove that

$$|h(x) - h(y)| \le \sum_{i: Q_i \cap \gamma \ne \emptyset} \rho(Q_i)$$

for points $x, y \in \gamma \cap S$, where $\rho(Q_i)$ is to be chosen. This will imply that the upper gradient inequality holds along the family of paths which are good for both ϕ and ψ, and thus for almost every path by the subadditivity of modulus.

If the endpoints x, y of γ lie in $S \cap V$, then we have

$$|h(x) - h(y)| \le \sum_{i: Q_i \cap \gamma \ne \emptyset} \operatorname*{osc}_{Q_i}(\phi),$$

and if $x, y \in S \setminus V$, then

$$|h(x) - h(y)| \le \sum_{i: Q_i \cap \gamma \ne \emptyset} \operatorname*{osc}_{Q_i}(\psi).$$

Now, suppose that $x \in S \cap V$ and $y \in S \setminus V$, but the path γ does *not* intersect $\partial_* V$. This implies that γ intersects some peripheral disk Q_{i_0} with $\partial Q_{i_0} \cap \partial V \ne \emptyset$. Indeed, consider the set

$$\widetilde{V} = V \cup \left(\bigcup_{i: \partial Q_i \subset V} Q_i \right),$$

which has the properties that $\widetilde{V} \cap S = V \cap S$, and $\partial_* \widetilde{V} = \partial_* V$. Note that γ intersects $\partial \widetilde{V}$ at a point z, since it has to exit $\widetilde{V}$. Furthermore, z cannot lie in S because $\gamma \cap$

$\partial_* V = \gamma \cap \partial_* \widetilde{V} = \emptyset$, but it has to lie in some peripheral disk Q_{i_0}. We assume that z is the first point of $\partial \widetilde{V}$ that γ hits as it travels from x to y. Let $x_0 \in \partial Q_{i_0} \cap \gamma$ be the first entry point of γ in ∂Q_{i_0}, and note that necessarily $x_0 \in \widetilde{V} \cap S = V \cap S$. Since $\partial Q_{i_0} \not\subset V$ (otherwise $z \in \widetilde{V}$), we have $\partial Q_{i_0} \cap \partial V \neq \emptyset$, so we fix a point $w \in \partial Q_{i_0} \cap \partial V \subset \partial_* V$. We now have

$$
\begin{aligned}
|h(x) - h(y)| &= |\phi(x) - \psi(y)| \\
&\leq |\phi(x) - \phi(x_0)| + |\phi(x_0) - \phi(w)| \\
&\quad + |\psi(w) - \psi(x_0)| + |\psi(x_0) - \psi(y)| \\
&\leq 2 \left(\sum_{i:Q_i \cap \gamma \neq \emptyset} \operatorname*{osc}_{Q_i}(\phi) + \sum_{i:Q_i \cap \gamma \neq \emptyset} \operatorname*{osc}_{Q_i}(\psi) \right),
\end{aligned}
$$

where we used that $\phi(w) = \psi(w)$ by the assumption that $\phi = \psi$ on $\partial_* V$.

Finally, we assume that $x \in S \cap V$ and $y \in S \setminus V$ and there exists a point $z \in \gamma \cap \partial_* V$. Here we have the estimate

$$
\begin{aligned}
|h(x) - h(y)| &\leq |h(x) - h(z)| + |h(z) - h(y)| \\
&= |\phi(x) - \phi(z)| + |\psi(z) - \psi(y)| \\
&\leq \sum_{i:Q_i \cap \gamma \neq \emptyset} \operatorname*{osc}_{Q_i}(\phi) + \sum_{i:Q_i \cap \gamma \neq \emptyset} \operatorname*{osc}_{Q_i}(\psi).
\end{aligned}
$$

Summarizing, we may choose $\rho(Q_i) = 2(\operatorname{osc}_{Q_i}(\phi) + \operatorname{osc}_{Q_i}(\psi))$ for $i \in \mathbb{N}$. This is clearly locally square-summable, since $\{\operatorname{osc}_{Q_i}(\phi)\}_{i \in \mathbb{N}}$ and $\{\operatorname{osc}_{Q_i}(\psi)\}_{i \in \mathbb{N}}$ are. The proof is complete. $\qquad\square$

Remark 2.5.29 It is very crucial in the preceding lemma that $\phi = \psi$ on $\partial_* V$, and we do not merely have $\phi(x) = \psi(x)$ for accessible points $x \in \partial_* V$. Indeed, one can construct square Sierpiński carpets for which the conclusion fails, if we use $\phi = 0$ and $\psi = 1$, and the "interface" $\partial_* V = \partial V \cap S$ is too small to be "seen" by carpet modulus; this is to say, that the curves passing through $\partial_* V$ have carpet modulus equal to 0. More precisely, suppose that h is a function equal to 0 in V and equal to 1 outside V. Then h has $\{\operatorname{osc}_{Q_i}(h)\}_{i \in \mathbb{N}}$ as an upper gradient because $\partial_* V$ is not intersected by almost every curve. If ∂Q_i intersects both V and $\Omega \setminus \overline{V}$, then $\operatorname{osc}_{Q_i}(h) = 1$. Hence, the very last inequality in the lemma is already violated. Moreover, if the preceding statement is true for infinitely many Q_i that are compactly contained in Ω, then $\operatorname{osc}_{Q_i}(h)$ is not locally square-summable. The next lemma has the stronger assumption that $\phi = \psi$ not only on almost every curve passing through $\partial_* V$, but actually on almost every curve in Ω.

For technical reasons, we also need the following modification of the previous lemma:

Lemma 2.5.30 *Let $V \subset \Omega$ be an open set. Let $\phi, \psi \in \mathcal{W}^{1,2}_{*,\mathrm{loc}}(S)$ and suppose that there exists a path family G in Ω that contains almost every path, such that $\phi(x) = \psi(x)$ for all points $x \in S$ that are accessible by paths of G. Then $h :=$ $\phi \chi_{S \cap V} + \psi \chi_{S \setminus V} \in \mathcal{W}^{1,2}_{*,\mathrm{loc}}(S)$.*
  Moreover, for each $i \in \mathbb{N}$ we have $\mathrm{osc}_{Q_i}(h) = \mathrm{osc}_{Q_i}(\phi)$ if $\partial Q_i \subset V$, $\mathrm{osc}_{Q_i}(h) = \mathrm{osc}_{Q_i}(\psi)$ if $\partial Q_i \subset \Omega \setminus \overline{V}$, and $\mathrm{osc}_{Q_i}(h) \leq \mathrm{osc}_{Q_i}(\phi) + \mathrm{osc}_{Q_i}(\psi)$ otherwise.

The statement that G contains almost every path is, by definition, equivalent to saying that its complement has carpet modulus equal to zero. By Lemma 2.5.18, the assumption of the lemma is equivalent to saying that ϕ and ψ have the same normalized version. The conclusion is essentially that no matter how one modifies a function within its equivalence class, it still remains in the Sobolev space.

Proof The proof is elementary so we skip some steps. Consider the curve family G_0 which contains all curves that are good for both ϕ and ψ, and are contained in G.

For a fixed $i \in \mathbb{N}$ consider points $z, w \in \partial Q_i$. Using Lemma 2.4.3, we may find a path $\gamma \in G$ and a point $x \in \partial Q_i$ that is accessible from γ. Therefore, $\phi(x) = \psi(x)$. Now, if $h(z) = \phi(z)$ and $h(w) = \psi(w)$, we have

$$|h(z) - h(w)| \leq |\phi(z) - \phi(x)| + |\psi(x) - \psi(w)| \leq \underset{Q_i}{\mathrm{osc}}(\phi) + \underset{Q_i}{\mathrm{osc}}(\psi).$$

This shows one of the claimed oscillation inequalities. The other cases are trivial.

For the upper gradient inequality, let $\gamma \in G$ be a curve and $x, y \in \gamma \cap S$. If x and y are accessible by γ, then by assumption $\phi(x) = \psi(x) = h(x)$ and $\phi(y) = \psi(y) = h(y)$, hence

$$|h(x) - h(y)| = |\phi(x) - \phi(y)| \leq \sum_{i : Q_i \cap \gamma \neq \emptyset} \underset{Q_i}{\mathrm{osc}}(\phi).$$

If $x \in \partial Q_{i_x}$ is non-accessible then we may consider the last exit point of γ from Q_{i_x} as it travels from x to y, in order to obtain an additional contribution $\mathrm{osc}_{Q_{i_x}}(h) \leq \mathrm{osc}_{Q_{i_x}}(\phi) + \mathrm{osc}_{Q_{i_x}}(\psi)$ in the above sum. The same comment applies if y is non-accessible. Thus, in all cases

$$|h(x) - h(y)| \leq \sum_{i : Q_i \cap \gamma \neq \emptyset} \left(\underset{Q_i}{\mathrm{osc}}(\phi) + \underset{Q_i}{\mathrm{osc}}(\psi) \right).$$

This shows that $\rho(Q_i) := \mathrm{osc}_{Q_i}(\phi) + \mathrm{osc}_{Q_i}(\psi)$ is an upper gradient of h. Using Remark 2.5.13 we derive the desired conclusion. $\square$

Finally, we need a special instance of Lemma 2.5.28:

Corollary 2.5.31 *Let $V \subset \Omega$ be an open set such that $\partial_* V \neq \emptyset$. Let $\psi \in \mathcal{W}^{1,2}_{*,\mathrm{loc}}(S)$ and $M \in \mathbb{R}$ be such that $\psi \leq M$ on $\partial_* V$. Then $h := (\psi \wedge M)\chi_{S \cap V} + \psi\chi_{S \setminus V} \in \mathcal{W}^{1,2}_{*,\mathrm{loc}}(S)$. Moreover, $\mathrm{osc}_{Q_i}(h) \leq \mathrm{osc}_{Q_i}(\psi)$ for all $i \in \mathbb{N}$.*

Proof The function $\phi := \psi \wedge M$ lies in $\mathcal{W}^{1,2}_{*,\mathrm{loc}}(S)$ by Lemma 2.5.27(c), with $\mathrm{osc}_{Q_i}(\phi) \leq \mathrm{osc}_{Q_i}(\psi)$. Since $\phi = \psi$ on $\partial_* V$, it follows that $h \in \mathcal{W}^{1,2}_{*,\mathrm{loc}}(S)$ by Lemma 2.5.28. It remains to show the oscillation inequality.

We fix $i \in \mathbb{N}$. If ∂Q_i is contained in V or in $S \setminus V$, then there is nothing to show, since $\mathrm{osc}_{Q_i}(\phi) \leq \mathrm{osc}_{Q_i}(\psi)$. Hence, we assume that ∂Q_i intersects ∂V, i.e., $\emptyset \neq \partial Q_i \cap \partial V \subset \partial_* V$. Using the assumption that $\psi \leq M$ on $\partial_* V$, we see that $m_{Q_i}(\psi) \leq M$. If $\psi|_{\partial Q_i \cap V} \leq M$, then there is nothing to show, since $h = \psi$ on ∂Q_i. Suppose that there exists $z \in \partial Q_i \cap V$ such that $\psi(z) > M$. Then

$$m_{Q_i}(\psi) \leq M < M_{Q_i}(\psi).$$

This implies that $\mathrm{osc}_{Q_i}(h) \leq \mathrm{osc}_{Q_i}(\psi)$, as desired.   □

2.6   Carpet-Harmonic Functions

Throughout the section we fix a relative Sierpiński carpet (S, Ω) with the standard assumptions.

2.6.1   Definition of Carpet-Harmonic Functions

Let $V \subset \Omega$ be an open set, and $f \in \mathcal{W}^{1,2}_{w,\mathrm{loc}}(S)$ (resp. $\mathcal{W}^{1,2}_{s,\mathrm{loc}}(S)$). Define the *(Dirichlet) energy functional* by

$$D_V(f) = \sum_{i \in I_V} \mathrm{osc}_{Q_i}(f)^2 \in [0, \infty].$$

Using the energy functional we define the notion of a weak (strong) carpet-harmonic function.

Definition 2.6.1 A function $u \in \mathcal{W}^{1,2}_{w,\mathrm{loc}}(S)$ (resp. $\mathcal{W}^{1,2}_{s,\mathrm{loc}}(S)$) is *weak* (resp. *strong*) *carpet-harmonic* if for every open set $V \subset\subset \Omega$ and each $\zeta \in \mathcal{W}^{1,2}_{w}(S)$ (resp. $\mathcal{W}^{1,2}_{s}(S)$) with $\zeta|_{S \setminus V} \equiv 0$ we have

$$D_V(u) \leq D_V(u + \zeta).$$

In other words, u minimizes the energy functional D_V over Sobolev functions with the same boundary values as u.

The functions u, ζ in the above definition are not assumed to be normalized, in the sense of the discussion in Sect. 2.5.2. Later we will see that the normalized version of a carpet-harmonic function has to be continuous; see Theorem 2.7.4.

For several statements in what follows there is no essential difference in using weak or strong carpet-harmonic functions. So, we will refer to them merely as carpet-harmonic functions and it will be implicitly understood that they lie in the corresponding Sobolev space. We will only specify that we are working with weak or strong carpet-harmonic functions if there is an actual difference.

Example 2.6.2 Let (S, Ω) be a relative Sierpiński carpet such that all peripheral disks Q_i are squares with sides parallel to the coordinate axes; see Fig. 2.4. Then the coordinate functions $u(x, y) = x$ and $v(x, y) = y$ are both weak and strong carpet-harmonic. We present a detailed proof below.

Since u, v are Lipschitz, Example 2.5.19 implies that they both lie in $\mathcal{W}^{1,2}_{s,\mathrm{loc}}(S) \subset \mathcal{W}^{1,2}_{w,\mathrm{loc}}(S)$.

Let $V \subset\subset \Omega$ be an open set and consider the open set $V' = V \cup \bigcup_{i \in I_V} Q_i \supset V$. This set contains all the peripheral disks that it intersects and it is also compactly contained in Ω. Moreover, V' satisfies $I_{V'} = I_V$, and thus $D_{V'} \equiv D_V$. We will show that $D_V(v) \leq D_V(g)$ for all $g \in \mathcal{W}^{1,2}_w(S) \supset \mathcal{W}^{1,2}_s(S)$ with $g = v$ outside V'. This suffices for harmonicity. Indeed, if $g \in \mathcal{W}^{1,2}_w(S)$ is arbitrary with $g = v$ outside V, then $g = v$ outside $V' \supset V$, so $D_V(v) \leq D_V(g)$, which shows harmonicity.

From now on, we denote V' by V and we will use the property that it contains the peripheral disks that it intersects. Let $g \in \mathcal{W}^{1,2}_w(S)$ with $g = v$ outside V. Note that for a.e. $x \in \mathbb{R}$ the vertical line γ_x passing through x (or rather its subpaths that lie in Ω) is a good path for g, by an argument very similar to the proof of

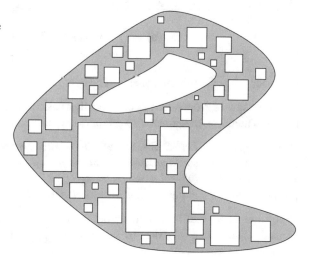

Fig. 2.4 A square relative Sierpiński carpet (S, Ω). Here Ω has two boundary components, the curves that are not squares

Lemma 2.4.3. We fix x such that γ_x is good and $\gamma_x \cap V \neq \emptyset$. The intersection is an open subset of a line, so it can be written as an (at most) countable union of disjoint open intervals $J_j = \{x\} \times (a_j, b_j)$, $j \in \mathbb{N}$. The points (x, a_j), (x, b_j) must lie in ∂V, and therefore, they lie in $\partial V \cap S$, by the assumption on V. By the fact that we have $g(x, y) = v(x, y) = y$ for $(x, y) \notin V$, together with the upper gradient inequality, we obtain

$$b_j - a_j = |g(x, b_j) - g(x, a_j)| \leq \sum_{i : Q_i \cap J_j \neq \emptyset} \operatorname*{osc}_{Q_i}(g).$$

Summing over all j and noting that a square Q_i cannot intersect two distinct sets J_j, we have

$$\mathcal{H}^1(\gamma_x \cap V) \leq \sum_{\substack{i : Q_i \cap \gamma_x \neq \emptyset \\ i \in I_V}} \operatorname*{osc}_{Q_i}(g).$$

Integrating over x and using Fubini's theorem in both sides, we have

$$\mathcal{H}^2(V) \leq \sum_{i \in I_V} \operatorname*{osc}_{Q_i}(g) \int_{Q_i \cap \gamma_x \neq \emptyset} dx = \sum_{i \in I_V} \operatorname*{osc}_{Q_i}(g) \ell(Q_i),$$

where $\ell(Q_i)$ is the side length of Q_i. Now, using the Cauchy-Schwarz inequality and the fact that $\mathcal{H}^2(V) = \sum_{i \in I_V} \ell(Q_i)^2$ (because $\mathcal{H}^2(S) = 0$), we see that

$$\sum_{i \in I_V} \ell(Q_i)^2 \leq \sum_{i \in I_V} \operatorname*{osc}_{Q_i}(g)^2.$$

On the other hand, it is easy to see that

$$D_V(v) = \sum_{i \in I_V} \operatorname*{osc}_{Q_i}(v)^2 = \sum_{i \in I_V} \ell(Q_i)^2,$$

so indeed v has the minimal energy. The computation for $u(x, y) = x$ is analogous.

As we saw in Proposition 2.5.24, locally quasiconformal maps preserve Sobolev spaces. Therefore, they must also preserve carpet-harmonic functions:

Proposition 2.6.3 *Let (S, Ω), (S', Ω') be relative Sierpiński carpets, and assume that $F : \Omega' \to \Omega$ is a locally quasiconformal map that maps the peripheral disks Q_i' of S' to the peripheral disks $Q_i = F(Q_i')$ of S. If $u : S \to \widehat{\mathbb{R}}$ is weak carpet-harmonic, then $u \circ F : S' \to \widehat{\mathbb{R}}$ is also weak carpet-harmonic.*

Proof Let $u : S \to \widehat{\mathbb{R}}$ be a weak carpet-harmonic function. Fix an open set $V' \subset\subset \Omega'$ and a function $\zeta' \in \mathcal{W}_w^{1,2}(S')$ such that $\zeta'|_{S' \setminus V'} \equiv 0$. Then $V := F(V')$ is compactly contained in Ω, and $\zeta := \zeta' \circ F^{-1} \in \mathcal{W}_w^{1,2}(S)$ with $\zeta|_{S \setminus V} \equiv 0$, by Proposition 2.5.24. Thus, by the correspondence of the peripheral disks and the harmonicity of u we have

$$D_{V'}(u \circ F + \zeta') = D_V(u \circ F \circ F^{-1} + \zeta' \circ F^{-1})$$

$$= D_V(u + \zeta) \geq D_V(u) = D_{V'}(u \circ F).$$

This shows that $u \circ F$ is weak carpet-harmonic, as desired. □

Corollary 2.6.4 *Let* (S, Ω), (S', Ω') *be relative Sierpiński carpets, and let* $F : S' \to S$ *be a local quasisymmetry. Furthermore, we assume that the peripheral circles of S and S' are uniform quasicircles. If $u : S \to \widehat{\mathbb{R}}$ is weak carpet-harmonic, then $u \circ F : S' \to \widehat{\mathbb{R}}$ is also weak carpet-harmonic.*

The proof follows immediately from the extension Lemma 2.5.26 and Proposition 2.6.3.

An interesting corollary of this discussion that relates carpet-harmonic functions to rigidity problems on square carpets is the following:

Corollary 2.6.5 *Let* (S, Ω), (S', Ω') *be relative Sierpiński carpets, and let* $F : S' \to S$ *be a local quasisymmetry. We assume that the peripheral circles of S' are uniform quasicirlces and that the peripheral circles of S are squares with sides parallel to the coordinate axes. Then the coordinates u, v of the map $F = (u, v)$ are weak carpet-harmonic.*

Proof By Example 2.6.2 the x, y-coordinate functions on S are weak carpet-harmonic. Corollary 2.6.4 implies that the pullbacks u, v of the coordinates are weak carpet-harmonic. □

2.6.2 Solution to the Dirichlet Problem

Let (S, Ω) be a relative Sierpiński carpet such that $\partial\Omega$ consists of finitely many disjoint Jordan curves (recall that a Jordan curve is a homeomorphic image of S^1). We fix a function $f \in \mathcal{W}_*^{1,2}(S)$. Then we can define the boundary values of f on points $x \in \partial\Omega$ that are accessible by a good path γ, using an analog of Lemma 2.5.6. Namely, we consider a good open path $\gamma \subset \Omega$ such that $\overline{\gamma} \cap \partial\Omega \neq \emptyset$, and for $x \in \overline{\gamma} \cap \partial\Omega$ we define

$$f(x) = \liminf_{\substack{Q_i \to x \\ Q_i \cap \gamma \neq \emptyset}} M_{Q_i}(f).$$

By a variant of Lemma 2.5.6 this definition does not depend on the path γ with $x \in \overline{\gamma}$. For the non-accessible points $x \in \partial\Omega$ we define $f(x) = \liminf_{y \to x} f(y)$ where $y \in \partial\Omega$ is accessible. Note that every point $x \in \partial\Omega$ is the landing point of a (not necessarily good) path $\gamma \subset \Omega$, by our assumptions on $\partial\Omega$. Perturbing γ as in Lemma 2.4.4 we obtain a point y near x that is accessible by a good path. Hence, there is a dense set of points in $\partial\Omega$ which are accessible by good paths, and this implies that the boundary values of f are well-defined on all of $\partial\Omega$.

We say that a function $u \in \mathcal{W}^{1,2}_*(S)$ *has boundary values equal to f* if there exists a path family $\mathcal{G}_0$ in Ω, whose complement has carpet modulus zero, such that $u(x) = f(x)$ for all points $x \in \partial\Omega$ that are accessible by a path $\gamma \in \mathcal{G}_0$.

Theorem 2.6.6 *Suppose that Ω is bounded and let $f \in \mathcal{W}^{1,2}_*(S)$ be a function with bounded boundary values, i.e., there exists $M > 0$ such that $|f(x)| \leq M$ for all $x \in \partial\Omega$. Then there exists a unique function $u \in \mathcal{W}^{1,2}_*(S)$ that minimizes $D_\Omega(g)$ over all $g \in \mathcal{W}^{1,2}_*(S)$ with boundary values equal to f. The function u is carpet-harmonic.*

The proof follows a general scheme of finding energy minimizers and extremal modulus densities in metric spaces. This scheme appears for example in [9, Proposition 2.4] in the carpet setting and also in [39] in the abstract metric space setting.

Proof The uniqueness part will be postponed until we have established several properties of carpet-harmonic functions; see Theorem 2.7.6. For the existence part, one has to minimize $D_\Omega(g)$ over all $g \in \mathcal{W}^{1,2}_*(S)$ with boundary values equal to f. We say that such a function g is *admissible (for the Dirichlet problem)*. It is easy to show that if g minimizes $D_\Omega(g)$ then it is carpet-harmonic. Indeed, for every $\zeta \in \mathcal{W}^{1,2}_*(S)$ that vanishes outside an open set $V \subset\subset \Omega$ we have

$$D_\Omega(g) \leq D_\Omega(g + \zeta).$$

Note that $\mathrm{osc}_{Q_i}(g + \zeta) = \mathrm{osc}_{Q_i}(g)$ for $i \notin I_V$. Canceling the common terms we obtain $D_V(g) \leq D_V(g + \zeta)$, so g is carpet-harmonic.

Define $D = \inf D_\Omega(g)$ where the infimum is taken over all admissible functions, and is finite since f is admissible. Let $g_n \in \mathcal{W}^{1,2}_*(S)$ be a minimizing sequence of admissible functions, i.e., $D_\Omega(g_n) \to D$ as $n \to \infty$. Note that $G_n := (g_n \wedge M) \vee (-M)$ is still a Sobolev function with $D_\Omega(G_n) \leq D_\Omega(g_n)$, by Proposition 2.5.27(b),(c), since M is a constant function. Thus, by replacing g_n with G_n, we may assume that $|g_n| \leq M$.

Now, we have a minimizing sequence g_n that satisfies $|M_{Q_i}(g_n)| \leq M$ for all $i \in \mathbb{N}$. In particular, the sequences $\{M_{Q_i}(g_n)\,\mathrm{diam}(Q_i)\}_{i\in\mathbb{N}}$, $\{\mathrm{osc}_{Q_i}(g_n)\}_{i\in\mathbb{N}}$ are uniformly bounded in $\ell^2(\mathbb{N})$. Here, it is crucial that $M_{Q_i}(g_n)\,\mathrm{diam}(Q_i) \leq M\,\mathrm{diam}(Q_i)$, and that

$$\sum_{i\in\mathbb{N}} \mathrm{diam}(Q_i)^2 < \infty$$

by the boundedness of Ω and the quasiroundness assumption (or by Corollary 2.3.6).

By passing to subsequences we may assume that for each $i \in \mathbb{N}$ we have $M_{Q_i}(g_n) \to M_{Q_i}$ and $\mathrm{osc}_{Q_i}(g_n) \to \rho(Q_i)$ for some real numbers $M_{Q_i}, \rho(Q_i)$. By Fatou's lemma we have

$$\sum_{i \in \mathbb{N}} \rho(Q_i)^2 \leq \liminf_{n \to \infty} \sum_{i \in \mathbb{N}} \mathrm{osc}_{Q_i}(g_n)^2 = D. \tag{2.30}$$

If we show that $\rho(Q_i)$ corresponds to the oscillation of an admissible function g, then this will be the desired minimizer.

Applying the Banach–Alaoglu theorem, we assume that $\{M_{Q_i}(g_n) \, \mathrm{diam}(Q_i)\}_{i \in \mathbb{N}}$ and $\{\mathrm{osc}_{Q_i}(g_n)\}_{i \in \mathbb{N}}$ converge weakly in $\ell^2(\mathbb{N})$, as $n \to \infty$, to $\{M_{Q_i} \, \mathrm{diam}(Q_i)\}_{i \in \mathbb{N}}$ and $\{\rho(Q_i)\}_{i \in \mathbb{N}}$, respectively. Since $\ell^2(\mathbb{N}) \times \ell^2(\mathbb{N})$ is reflexive, by Mazur's lemma (see e.g. [47, Theorem 2, p. 120]) there exist convex combinations

$$M_{Q_i}^n := \sum_{j=1}^{n} \lambda_j^n M_{Q_i}(g_j), \quad \rho^n(Q_i) := \sum_{j=1}^{n} \lambda_j^n \mathrm{osc}_{Q_i}(g_j)$$

such that the sequences $\{M_{Q_i}^n \, \mathrm{diam}(Q_i)\}_{i \in \mathbb{N}}$ and $\{\rho^n(Q_i)\}_{i \in \mathbb{N}}$ converge strongly in $\ell^2(\mathbb{N})$ to $\{M_{Q_i} \, \mathrm{diam}(Q_i)\}_{i \in \mathbb{N}}$ and $\{\rho(Q_i)\}_{i \in \mathbb{N}}$, respectively.

We now show that $\{M_{Q_i}\}_{i \in \mathbb{N}}$ defines a discrete Sobolev function in the sense of Definition 2.5.1 with upper gradient $\{\rho(Q_i)\}_{i \in \mathbb{N}}$. Consider $\mathcal{G}$ to be the family of curves that are good curves for the functions $g_n, n \in \mathbb{N}$, and f, and also the boundary values of g_n along paths $\gamma \in \mathcal{G}$ are equal to f for $n \in \mathbb{N}$. Moreover, we assume that the paths of $\mathcal{G}$ are non-exceptional for Fuglede's Lemma 2.3.8, applied to a subsequence of $\{\rho^n(Q_i)\}_{i \in \mathbb{N}}$, which we still denote in the same way. For peripheral disks Q_{i_1}, Q_{i_2} that intersect a curve $\gamma \in \mathcal{G}$ we have

$$|M_{Q_{i_1}}^n - M_{Q_{i_2}}^n| \leq \sum_{j=1}^{n} \lambda_j^n |M_{Q_{i_1}}(g_j) - M_{Q_{i_2}}(g_j)|$$

$$\leq \sum_{j=1}^{n} \lambda_j^n \sum_{i : Q_i \cap \gamma \neq \emptyset} \mathrm{osc}_{Q_i}(g_j) \tag{2.31}$$

$$= \sum_{i : Q_i \cap \gamma \neq \emptyset} \rho^n(Q_i).$$

Taking limits we obtain

$$|M_{Q_{i_1}} - M_{Q_{i_2}}| \leq \sum_{i : Q_i \cap \gamma \neq \emptyset} \rho(Q_i).$$

Hence, $\{M_{Q_i}\}_{i \in \mathbb{N}}$ is indeed a discrete Sobolev function having $\{\rho(Q_i)\}_{i \in \mathbb{N}}$ as an upper gradient.

By the discussion in Sect. 2.5.1, the discrete Sobolev function $\{M_{Q_i}\}_{i\in\mathbb{N}}$ yields a Sobolev function g with upper gradient $\{\text{osc}_{Q_i}(g)\}_{i\in\mathbb{N}}$ that satisfies $\text{osc}_{Q_i}(g) \leq \rho(Q_i)$ for all $i \in \mathbb{N}$; see Corollary 2.5.7. Combining this with (2.30), we see that $D_\Omega(g) \leq D$. If we prove that g is admissible for the Dirichlet problem, then we will have

$$D_\Omega(g) = \sum_{i\in\mathbb{N}} \text{osc}_{Q_i}(g)^2 = \sum_{i\in\mathbb{N}} \rho(Q_i)^2 = D,$$

and in particular $\text{osc}_{Q_i}(g) = \rho(Q_i)$ for all $i \in \mathbb{N}$.

It remains to show that g is admissible for the Dirichlet problem. For this, it suffices to show that there exists a path family $\mathcal{G}_0$ that contains almost every path, such that for all points $x \in \partial\Omega$ which are accessible by paths in $\mathcal{G}_0$ we have $g(x) = f(x)$. Let $\mathcal{G}_0$ be the path family that contains all paths $\gamma \in \mathcal{G}$ for which

$$\sum_{i:Q_i\cap\gamma\neq\emptyset} \rho(Q_i) < \infty.$$

Note that the complement of $\mathcal{G}_0$ has carpet modulus zero. If $x \in \partial\Omega$ is accessible by a path $\gamma \in \mathcal{G}_0$, and Q_{i_1}, Q_{i_2} intersect γ, by (2.31) we have

$$|M_{Q_{i_1}}^n - M_{Q_{i_2}}^n| \leq \sum_{i:Q_i\cap\gamma\neq\emptyset} \rho^n(Q_i).$$

As we let $Q_{i_2} \to x$, the quantity $M_{Q_{i_2}}^n = \sum_{j=1}^n \lambda_j^n M_{Q_{i_2}}(g_j)$ converges to $f(x)$, because each term $M_{Q_{i_2}}(g_j)$ does so (recall that γ is non-exceptional for each g_j, and they are admissible). Hence, we have

$$|M_{Q_{i_1}}^n - f(x)| \leq \sum_{i:Q_i\cap\gamma\neq\emptyset} \rho^n(Q_i).$$

Now we let $n \to \infty$, and using Fuglede's Lemma 2.3.8 we obtain

$$|M_{Q_{i_1}} - f(x)| \leq \sum_{i:Q_i\cap\gamma\neq\emptyset} \rho(Q_i).$$

We claim that $M_{Q_{i_1}} \to g(x)$ as $Q_{i_1} \to x$ and $Q_{i_1} \cap \gamma \neq \emptyset$. We assume this for the moment, and we have

$$|g(x) - f(x)| \leq \sum_{i:Q_i\cap\gamma\neq\emptyset} \rho(Q_i).$$

This is also true for all subpaths of γ landing at x, so if we shrink γ to x, we obtain $g(x) = f(x)$, as desired.

To prove our claim, we note that, by the definition of boundary values, $g(x)$ can be approximated by $M_{Q_{i_1}}(g) = \sup_{x \in \partial Q_{i_1}}(g)$, as $Q_{i_1} \to x$ and $Q_{i_1} \cap \gamma \neq \emptyset$. On the other hand, by the last inequality in Corollary 2.5.7, we have

$$|M_{Q_{i_1}}(g) - M_{Q_{i_1}}| \leq \rho(Q_{i_1}).$$

Since ρ is square-summable, as $Q_{i_1} \to x$ we have $\rho(Q_{i_1}) \to 0$. Hence, our claim is proved. □

Remark 2.6.7 By the discussion in Sect. 2.5.2, there exists a normalized version $\tilde{g}$ of the solution g to the Dirichlet problem. For the normalized version we have $\mathrm{osc}_{Q_i}(\tilde{g}) \leq \mathrm{osc}_{Q_i}(g)$ for all $i \in \mathbb{N}$ by Lemma 2.5.16(iii). Hence, $D_\Omega(\tilde{g}) \leq D_\Omega(g)$. If $\tilde{g}$ has boundary values equal to f, then we will have that $\tilde{g}$ is admissible, hence $D_\Omega(\tilde{g}) = D_\Omega(g)$, and $\tilde{g}$ is also a solution to the Dirichlet problem.

However, $\tilde{g}$ agrees with g at all points which are accessible by paths $\gamma \in G_g$ by Lemma 2.5.16(ii). This also holds for accessible boundary points. Hence, indeed $\tilde{g}$ has boundary values equal to f.

2.7 Properties of Harmonic Functions

We remind the reader that we have dropped the terminology weak/strong for the Sobolev functions and carpet-harmonic functions. We also drop the subscripts w, s for the Sobolev spaces, so e.g., the non-local Sobolev space is denoted by $\mathcal{W}_*^{1,2}(S)$. All results below apply to both weak and strong carpet-harmonic functions. First we record a lemma that allows us to switch to the normalized version (see Sect. 2.5.2) of a harmonic function:

Lemma 2.7.1 *Let $u : S \to \widehat{\mathbb{R}}$ be a carpet-harmonic function. Then its normalized version $\tilde{u}$ is also carpet-harmonic.*

Proof It suffices to prove that for each open set $V \subset\subset \Omega$ and each function $\zeta \in \mathcal{W}_*^{1,2}(S)$ with $\zeta|_{S \setminus V} \equiv 0$ we have $D_V(\tilde{u}) \leq D_V(\tilde{u} + \zeta)$. We fix such a function ζ. Recall that $\mathrm{osc}_{Q_i}(\tilde{u}) \leq \mathrm{osc}_{Q_i}(u)$ for all $i \in \mathbb{N}$, by Lemma 2.5.16(iii). Hence, $D_V(\tilde{u}) \leq D_V(u)$. On the other hand, by Lemma 2.5.16(ii) we have $\tilde{u}(x) = u(x)$ for all points $x \in S$ that are accessible by a curve family that contains almost every curve. Hence, by Lemma 2.5.30 and linearity, for each open set $W \subset \Omega$ the function $\eta = (\tilde{u} - u)\chi_{S \cap W} + \zeta$ lies in the Sobolev space $\mathcal{W}_*^{1,2}(S)$.

First, assume that $\partial Q_i \subset V$ whenever $i \in I_V$. We set $W = V$ and consider the function η as above. Then $u + \eta = \tilde{u} + \zeta$ on $\partial Q_i \subset V$ for $i \in I_V$. Since η vanishes outside V, by the harmonicity of u we have

$$D_V(u) \leq D_V(u + \eta) = D_V(\tilde{u} + \zeta).$$

Summarizing, $D_V(\tilde{u}) \leq D_V(\tilde{u} + \zeta)$, as desired.

Now, we treat the general case. We fix $\varepsilon > 0$ and for each $i \in I_V$ we consider a number $\delta_i = \delta_i(\varepsilon) \in (0, \varepsilon)$ such that the open δ_i-neighborhood of Q_i intersects only peripheral disks having smaller diameter than that of Q_i; recall from Lemma 2.3.4 that the diameters of the peripheral disks shrink to 0 in compact subsets of $\mathbb{R}^2$. We denote by W_ε the union of V with all these neighborhoods and η is defined as before with $W = W_\varepsilon$. Note that W_ε contains ∂Q_i, whenever $i \in I_V$. Therefore, as η vanishes outside $W_\varepsilon \supset V$, we have

$$
\begin{aligned}
D_V(u) \leq D_{W_\varepsilon}(u) &\leq D_{W_\varepsilon}(u + \eta) \\
&= D_{W_\varepsilon}(\tilde{u}\chi_{S \cap W_\varepsilon} + u\chi_{S \setminus W_\varepsilon} + \zeta) \\
&= D_V(\tilde{u} + \zeta) + \sum_{i \in I_{W_\varepsilon} \setminus I_V} \operatorname*{osc}_{Q_i}(\tilde{u}\chi_{S \cap W_\varepsilon} + u\chi_{S \setminus W_\varepsilon})^2 \\
&\leq D_V(\tilde{u} + \zeta) + \sum_{i \in I_{W_\varepsilon} \setminus I_V} (\operatorname*{osc}_{Q_i}(\tilde{u}) + \operatorname*{osc}_{Q_i}(u))^2,
\end{aligned}
$$

where we used the oscillation inequalities from Lemma 2.5.30. Since the last sum is finite, it will converge to 0 and we will have the desired conclusion, provided that the set I_{W_ε} shrinks to I_V as $\varepsilon \to 0$.

We argue by contradiction, assuming that there exists $i_0 \notin I_V$ that lies in I_{W_ε} infinitely often as $\varepsilon \to 0$, say along a sequence $\varepsilon_n \to 0$. For each $n \in \mathbb{N}$, there exists $i(n) \in I_V$ such that Q_{i_0} intersects the $\delta_{i(n)}(\varepsilon_n)$-neighborhood of $Q_{i(n)}$. Note that the set $\{i(n) : n \in \mathbb{N}\}$ cannot be finite, since $\delta_{i(n)}(\varepsilon_n) \to 0$ as $n \to \infty$, and Q_{i_0} has positive distance from each individual peripheral disk. Therefore, the set $\{i(n) : n \in \mathbb{N}\}$ is infinite, and by passing to a subsequence we may assume that the indices $i(n)$ are distinct. However, $\delta_{i(n)}(\varepsilon_n)$ was chosen so that the $\delta_{i(n)}(\varepsilon_n)$-neighborhood of $Q_{i(n)}$ intersects only smaller peripheral disks than $Q_{i(n)}$. Since $\operatorname{diam}(Q_{i(n)}) \to 0$ (by Lemma 2.3.4 since they all stay near Q_{i_0}), it follows that Q_{i_0} cannot intersect these neighborhoods for large n. This is a contradiction.   □

Recall also by Remark 2.6.7 that the normalized version of the solution to the Dirichlet problem (Theorem 2.6.6) has the same boundary values as the original solution. In what follows, we always use normalized versions of carpet-harmonic functions. In particular, by Lemma 2.5.17 we may assume that for any $x \in S$ the value of $u(x)$ can be approximated by $M_{Q_i}(u)$ where Q_i is a peripheral disk close to x.

2.7.1 Continuity and Maximum Principle

Lemma 2.7.2 *Let* $u \colon S \to \widehat{\mathbb{R}}$ *be a normalized Sobolev function, and* $V \subset \Omega$ *a connected open set.*

(a) *If* $\operatorname{osc}_{Q_i}(u) = 0$ *for all* $i \in I_V$, *then* u *is constant on* $S \cap V$.
(b) *If* $\operatorname{osc}_{Q_i}(u) = 0$ *for all* $Q_i \subset V$, *then* u *is a constant on* $S \cap W$, *where* W *is any component of* $V \setminus \overline{\bigcup_{i \in I_{\partial V}} Q_i}$.

Proof (a) Note that u is constant on (the boundary of) any given peripheral disk that intersects V. Thus we may assume that V does not intersect only one peripheral disk. Let $i_1, i_2 \in I_V$ be distinct. Since V is path connected, by Lemma 2.4.3 we can find a non-exceptional path $\gamma \subset V$ joining Q_{i_1}, Q_{i_2} such that the upper gradient inequality holds along γ. If $x \in \partial Q_{i_1} \cap \gamma$ and $y \in \partial Q_{i_2} \cap \gamma$ then

$$|u(x) - u(y)| \leq \sum_{i : Q_i \cap \gamma \neq \emptyset} \operatorname*{osc}_{Q_i}(u) = 0.$$

Thus u is equal to the same constant c on all peripheral circles ∂Q_i, $i \in I_V$. If $x \in S \cap V$ is arbitrary, then $u(x)$ can be approximated by $M_{Q_i}(u)$, where $Q_i \cap V \neq \emptyset$, thus $u(x) = c$ also here.

(b) The set $V \setminus \overline{\bigcup_{i \in I_{\partial V}} Q_i}$ is open. We fix a component W of this set and observe that if $Q_i \cap W \neq \emptyset$ then $Q_i \subset W$; see the remark below. This implies that $I_W = \{i \in \mathbb{N} : Q_i \subset W\}$. Thus the conclusion follows immediately by an application of part (a) of the lemma. $\qquad\square$

Remark 2.7.3 (a) Each component W of $V \setminus \overline{\bigcup_{i \in I_{\partial V}} Q_i}$ has the property that it contains all peripheral disks that it intersects. Moreover, if $S \cap W \neq \emptyset$ and $\partial_* V \neq \emptyset$, then we have $\overline{S \cap W} \cap \partial_* V \neq \emptyset$.

For the first claim, note that $E := \overline{\bigcup_{i \in I_{\partial V}} Q_i}$ contains all the peripheral disks that it intersects. This implies that if $Q_i \cap W \neq \emptyset$, then we necessarily have $Q_i \cap E = \emptyset$ and $Q_i \cap \partial V = \emptyset$. Hence, $Q_i \subset V \setminus E$ and $Q_i \subset W$ by the connectedness of Q_i.

For the second claim, by Lemma 2.4.10, there exists an open path $\gamma \subset S^\circ$ connecting a point x of $S \cap W$ to a point outside V, for example to a point of $\partial_* V$. Let $y \in \gamma \cap \partial V$ be the first point of ∂V that γ meets, assuming that it is parametrized to start at x. We claim that $y \in \overline{S \cap W}$. We consider the (smallest) open subpath of γ that connects x to y, and we still denote it by γ. Then $\gamma \subset V$, and also $\gamma \cap \bigcup_{i \in I_{\partial V}} Q_i = \emptyset$. Indeed, if $z \in \gamma \subset S^\circ$ is a limit point of $\bigcup_{i \in I_{\partial V}} Q_i$, then there exists a sequence of Q_i, $i \in I_{\partial V}$, with diameters shrinking to 0 and with $Q_i \to z$. This would imply that $z \in \partial V$, a contradiction. Hence, $\gamma \subset V \setminus \bigcup_{i \in I_{\partial V}} Q_i$, and in fact $\gamma \subset S \cap W$, which shows that $y \in \overline{\gamma} \subset \overline{S \cap W}$.

(b) The assumption $\partial_* V \neq \emptyset$ in the previous statement holds always, unless $V \supset S$ or $\mathbb{C} \setminus \overline{V} \supset S$. Indeed, if V is an open set and $S \setminus V \neq \emptyset$, $S \cap \overline{V} \neq \emptyset$, then we can connect a point of $S \cap \overline{V}$ to a point of $S \setminus V$ with an open path in S°, by Lemma 2.4.10. This path necessarily intersects $\partial V \cap S = \partial_* V$.

Theorem 2.7.4 *Let* $u : S \to \widehat{\mathbb{R}}$ *be a carpet-harmonic function. Then* u *is continuous.*

Proof Let $x \in S^\circ$. If $\mathrm{osc}_{Q_i}(u) = 0$ for all Q_i contained in a ball $B(x, r)$ then there exists $r' < r$ such that $\mathrm{osc}_{Q_i}(u) = 0$ for all Q_i intersecting the ball $B(x, r')$. This follows from Lemma 2.3.4 and the fact that no peripheral disk can intersect a ball $B(x, r')$ for arbitrarily small $r' > 0$. Applying the previous lemma, we conclude that u is constant in $B(x, r') \cap S$, so it is trivially continuous.

We assume that arbitrarily close to x there exists some Q_i with $\mathrm{osc}_{Q_i}(u) > 0$. Consider a circular arc γ_r as in Lemma 2.4.7(a) with $\sum_{i:Q_i \cap \gamma_r \neq \emptyset} \mathrm{osc}_{Q_i}(u) < \varepsilon$ and $B(x, r) \subset \Omega$. Also, let $y \in B(x, r) \cap S$. Since u is normalized, there exist peripheral disks $Q_{i_x}, Q_{i_y} \subset B(x, r)$ such that $|u(x) - M_{Q_{i_x}}(u)| < \varepsilon$ and $|u(y) - M_{Q_{i_y}}(u)| < \varepsilon$, so it suffices to show that $|M_{Q_{i_x}}(u) - M_{Q_{i_y}}(u)|$ is small.

Since γ_r is non-exceptional, the upper gradient inequality implies that the number $M := \sup_{z \in \partial_* B(x,r)} u(z) = \sup_{z \in S \cap \gamma_r} u(z)$ is finite. We claim that $M_{Q_k}(u) \leq M$, for all $Q_k \subset B(x, r)$. Consider the function $h = u \cdot \chi_{S \backslash B(x,r)} + u \wedge M \cdot \chi_{S \cap B(x,r)}$. By Corollary 2.5.31, it follows that h is a Sobolev function and $\mathrm{osc}_{Q_i}(h) \leq \mathrm{osc}_{Q_i}(u)$ for all $i \in \mathbb{N}$. Therefore, for the Dirichlet energy we have $D_{B(x,r)}(h) \leq D_{B(x,r)}(u)$.

Assume now that there exists some $Q_k \subset B(x, r)$ with $M_{Q_k}(u) > M$. If $\mathrm{osc}_{Q_k}(u) > 0$, then it is easy to see that $\mathrm{osc}_{Q_k}(h) < \mathrm{osc}_{Q_k}(u)$, which implies that $D_{B(x,r)}(h) < D_{B(x,r)}(u)$, a contradiction to harmonicity. If $\mathrm{osc}_{Q_k}(u) = 0$, then consider a good path $\gamma \subset B(x, r)$, given by Lemma 2.4.3, that joins Q_k to some $Q_l \subset B(x, r)$ with $\mathrm{osc}_{Q_l}(u) > 0$. Using the upper gradient inequality one can then find a peripheral disk $Q_m \subset B(x, r)$ that intersects γ, such that $\mathrm{osc}_{Q_m}(u) > 0$ and $M_{Q_m}(u)$ is arbitrarily close to $M_{Q_k}(u)$, so, in particular, $M_{Q_m}(u) > M$. By the previous case we obtain a contradiction.

With a similar argument, one shows that $\inf_{z \in \partial_* B(x,r)} u(z) \leq M_{Q_k}(u)$ for all $Q_k \subset B(x, r)$. Therefore by the upper gradient inequality we have

$$|M_{Q_{i_x}}(u) - M_{Q_{i_y}}(u)| \leq \sup_{z \in \partial_* B(x,r)} u(z) - \inf_{z \in \partial_* B(x,r)} u(z)$$

$$\leq \sum_{i:Q_i \cap \gamma_r \neq \emptyset} \mathrm{osc}_{Q_i}(u) < \varepsilon,$$

as desired.

Now, we treat the case $x \in \partial Q_{i_0}$ for some $i_0 \in \mathbb{N}$. Consider a small ball $B(x, r)$ with $\partial B(x, r) \cap Q_{i_0} \neq \emptyset$. If $\mathrm{osc}_{Q_i}(u) = 0$ for all Q_i contained in $B := B(x, r)$, then from Lemma 2.7.4(b) for the component W of $B(x, r) \backslash \overline{\bigcup_{i \in I_{\partial B}} Q_i}$ that contains x in its boundary we have that $u|_{S \cap W}$ is a constant c. In fact $\partial Q_{i_0} \cap \partial W$ contains a non-trivial arc α that contains x in its interior. Since u is normalized, Lemma 2.5.17 implies that the value of $u(y)$ for $y \in \alpha$ is approximated by $M_{Q_i}(u) = c$, where $Q_i \subset W$. This shows that $u \equiv c$ in a neighborhood of x, and thus, u is continuous at x.

Now, we assume that arbitrarily close to x there exists some Q_i, $i \neq i_0$, with $\mathrm{osc}_{Q_i}(u) > 0$. For a small $\varepsilon > 0$ we apply again Lemma 2.4.7(a) to obtain a circular arc γ_r around x such that

$$\sum_{\substack{i:Q_i\cap\gamma_r\neq\emptyset \\ i\neq i_0}} \mathrm{osc}_{Q_i}(u) < \varepsilon.$$

As in the proof of Lemma 2.5.6 (see also Fig. 2.2), there exists a (closed) subarc γ_r' of γ_r with endpoints on ∂Q_{i_0} such that $\gamma_r' \cap Q_{i_0} = \emptyset$ and γ_r' defines a crosscut that separates x from ∞ in $\mathbb{R}^2 \setminus Q_{i_0}$. We consider an arc $\beta \subset \overline{Q}_{i_0}$ whose endpoints are the endpoints of γ_r', but otherwise it is contained in Q_{i_0}. Then $\beta \cup \gamma_r'$ bounds a Jordan region V that contains x in its interior.

With a similar variational argument as in the case $x \in S^\circ$ we will show that for each $Q_k \subset V$ we have

$$\inf_{z\in S\cap\gamma_r'} u(z) \leq M_{Q_k}(u) \leq \sup_{z\in S\cap\gamma_r'} u(z).$$

Then continuity will follow as before, because $\sum_{i:Q_i\cap\gamma_r'\neq\emptyset} \mathrm{osc}(Q_i) < \varepsilon$.

We sketch the proof of the right inequality. Let $M = \sup_{z\in S\cap\gamma_r'} u(z)$ (which is finite by the upper gradient inequality for the good path γ_r'), and consider the function $h = u \cdot \chi_{S\setminus V} + u \wedge M \cdot \chi_{S\cap V}$. By Corollary 2.5.31 the function h is a Sobolev function which agrees with u outside V and $\mathrm{osc}_{Q_i}(h) \leq \mathrm{osc}_{Q_i}(u)$ for all $i \in \mathbb{N}$. Now, if there exists $Q_k \subset V$ with $M_{Q_k}(u) > M$ we derive a contradiction as in the previous case.   □

The continuity implies, in particular, that $|u(x)| < \infty$ for every $x \in S$.

Theorem 2.7.5 (Maximum Principle) *Let $u: S \to \mathbb{R}$ be a carpet-harmonic function and $V \subset\subset \Omega$ be an open set. Then*

$$\sup_{x\in S\cap\overline{V}} u(x) = \sup_{x\in\partial_* V} u(x) \quad and \quad \inf_{x\in S\cap\overline{V}} u(x) = \inf_{x\in\partial_* V} u(x).$$

Note that if $S \cap \overline{V} \neq \emptyset$, then $\partial_* V \neq \emptyset$. This is because $V \subset\subset \Omega$; recall Remark 2.7.3(b). Hence, the sets under the suprema and infima are simultaneously empty or simultaneously non-empty. Moreover, by the continuity of u all quantities are finite.

Proof We clearly have $\sup_{x\in S\cap\overline{V}} u(x) \geq \sup_{x\in\partial_* V} u(x) =: M$ because $S \cap \overline{V} \supset \partial_* V$.

Assume that there exists $x \in S \cap V$ such that $u(x) > M$. Since S° is dense in S (this follows e.g. by Lemma 2.4.10) and u is continuous, we may assume that there exists $x \in S^\circ \cap V$ such that $u(x) > M$. Note that x cannot lie in $\overline{\bigcup_{i\in I_{\partial V}} Q_i}$ since all peripheral disks contained in a small neighborhood of x have to lie in V. Let W

be the component of $V \setminus \overline{\bigcup_{i \in I_{\partial V}} Q_i}$ that contains x. By Remark 2.7.3(a), we have
$\overline{S \cap W} \cap \partial_* V \neq \emptyset$ and W contains all peripheral disks that it intersects.

If $\mathrm{osc}_{Q_i}(u) = 0$ for all $Q_i \subset W$, then u is constant in $S \cap W$ by Lemma 2.7.2(a)
and by continuity it is also constant and equal to $u(x)$ on $\overline{S \cap W}$. This contradicts
the fact that $\overline{S \cap W} \cap \partial_* V \neq \emptyset$, and that $u \leq M$ on $\partial_* V$.

Hence, there exists some $Q_i \subset W$ with $\mathrm{osc}_{Q_i}(u) > 0$. Arbitrarily close to x
we can find a peripheral disk Q_{i_x} with $M_{Q_{i_x}}(u) > M$. Arguing as in the proof of
Theorem 2.7.4, we can derive that there exists some $Q_{i_0} \subset W$ with $M_{Q_{i_0}}(u) > M$
and $\mathrm{osc}_{Q_{i_0}}(u) > 0$. Consider the variation $h = u \cdot \chi_{S \setminus V} + u \wedge M \cdot \chi_{S \cap V}$ and note
that $u \leq M$ on $\partial_* V$. By Corollary 2.5.31 h is a Sobolev function with $\mathrm{osc}_{Q_i}(h) \leq$
$\mathrm{osc}_{Q_i}(u)$ for all $i \in \mathbb{N}$. However $\mathrm{osc}_{Q_{i_0}}(h) < \mathrm{osc}_{Q_{i_0}}(u)$ which contradicts the
minimizing property of u.

The claim for the infimum follows by looking at $-u$.   □

2.7.2 Uniqueness and Comparison Principle

Here, we establish first the uniqueness part in Theorem 2.6.6, and then a comparison
principle for solutions to the Dirichlet problem. The standing assumption here is that
the set Ω has boundary $\partial \Omega$ that consists of finitely many disjoint Jordan curves, so
that we can define boundary values of Sobolev functions.

Theorem 2.7.6 (Uniqueness) *Let $u, v \colon S \to \mathbb{R}$ be solutions to the Dirichlet
problem given by Theorem 2.6.6 with boundary values equal to f on $\partial \Omega$. Then
$u = v$ on S.*

The proof of uniqueness contains ideas from the proof of [13, Theorem 7.14].

Proof Since both u, v are solutions to the Dirichlet problem, it follows that
$D_\Omega(u) = D_\Omega(v)$ and we define D to be this common number. Recall that a function
g is admissible for the Dirichlet problem if $g \in \mathcal{W}^{1,2}_*(S)$ and g has boundary values
equal to f.

For $s \in [0, 1]$ the function $(1 - s)u + sv$ is admissible, thus by the subadditivity
of $\mathrm{osc}_{Q_i}(\cdot)$ (see e.g. the proof of Proposition 2.5.27) and the Cauchy–Schwarz
inequality we have

$$
\begin{aligned}
D &\leq D_\Omega((1 - s)u + sv) \\
&\leq (1 - s)^2 D_\Omega(u) + 2s(1 - s) \sum_{i \in \mathbb{N}} \mathrm{osc}_{Q_i}(u)\, \mathrm{osc}_{Q_i}(v) + s^2 D_\Omega(v) \\
&\leq (1 - s)^2 D + 2s(1 - s) D^{1/2} D^{1/2} + s^2 D \\
&= D.
\end{aligned}
$$

Since we have equality, it follows that $\mathrm{osc}_{Q_i}(u) = \mathrm{osc}_{Q_i}(v)$ for all $i \in \mathbb{N}$.

Consider the function $g = u \vee v$, which is a Sobolev function with $\mathrm{osc}_{Q_i}(g) \leq \mathrm{osc}_{Q_i}(u)$, by Proposition 2.5.27(b). Also, $g(x) = f(x)$ for all accessible points $x \in \partial\Omega$, thus g is admissible for the Dirichlet problem with boundary data f. It follows that

$$D \leq D_\Omega(g) = \sum_{i \in \mathbb{N}} \mathrm{osc}_{Q_i}(g)^2 \leq \sum_{i \in \mathbb{N}} \mathrm{osc}_{Q_i}(u)^2 = D.$$

This implies that $\mathrm{osc}_{Q_i}(g) = \mathrm{osc}_{Q_i}(u)$ for all $i \in \mathbb{N}$, and g is also carpet-harmonic on Ω, since it minimizes the Dirichlet energy.

We assume that there exists $x_0 \in \Omega$ such that $u(x_0) < v(x_0)$. Then $u(x_0) < g(x_0)$, and there exists a level $\Lambda \in \mathbb{R}$ such that $u(x_0) < \Lambda < g(x_0)$. We define the function

$$h = \begin{cases} g, & g \leq \Lambda \\ \Lambda, & u < \Lambda < g \\ u, & u \geq \Lambda. \end{cases}$$

It is immediate to see that $h = (u \vee \Lambda) \wedge g$. Proposition 2.5.27(b),(c) shows that h is a Sobolev function with $\mathrm{osc}_{Q_i}(h) \leq \mathrm{osc}_{Q_i}(u) = \mathrm{osc}_{Q_i}(g)$ for all $i \in \mathbb{N}$. It is also clear that h is admissible for the Dirichlet problem on Ω with boundary data equal to f. It thus follows that h is also carpet-harmonic and in fact $\mathrm{osc}_{Q_i}(h) = \mathrm{osc}_{Q_i}(u)$ for all $i \in \mathbb{N}$.

If the closure of $\{u < \Lambda < g\}$ relative to S is the entire carpet S, then $h \equiv \Lambda$ is constant and $\mathrm{osc}_{Q_i}(h) = \mathrm{osc}_{Q_i}(u) = \mathrm{osc}_{Q_i}(v) = 0$. Lemma 2.7.2(a) implies that u, v are constants, but then they cannot have the same boundary values, unless $u \equiv v$. This contradicts the assumption that $u(x_0) < v(x_0)$. Hence, we assume that there exists a point of S that does not lie in $\{u < \Lambda < g\}$.

We will show that there exists a peripheral disk $Q_{i_0} \subset Z := \{u < \Lambda < g\}$ with $\mathrm{osc}_{Q_{i_0}}(u) > 0$. However, h is constant in $S \cap Z$, thus $\mathrm{osc}_{Q_{i_0}}(h) = 0$, which is again a contradiction, because $\mathrm{osc}_{Q_{i_0}}(h) = \mathrm{osc}_{Q_{i_0}}(u)$.

To prove our claim, note first that by the continuity of the carpet-harmonic functions u, g the set Z is the intersection of an open set V in the plane with S, and Z is non-empty, since it contains x_0. Since $S \setminus V = S \setminus Z \neq \emptyset$ and $S \cap V \neq \emptyset$, we have $\partial_* V \neq \emptyset$; see Remark 2.7.3(b). Let W be the component of $V \setminus \bigcup_{i \in I_{\partial V}} Q_i$ that contains x_0. Then $\overline{S \cap W} \cap \partial_* V \neq \emptyset$, by Remark 2.7.3(a). If $\mathrm{osc}_{Q_i}(u) = 0$ for all $Q_i \subset W$, then u is constant in $S \cap W$ by Lemma 2.7.2(a) and by continuity it is also constant on $\overline{S \cap W} \cap \partial_* V$. In particular, there exists a point $z_0 \in \partial_* V \subset \partial V$ with $u(z_0) = u(x_0) < \Lambda$ and $g(z_0) > \Lambda$. Since these inequalities hold in a neighborhood of z_0 we obtain a contradiction. $\qquad\square$

Theorem 2.7.7 (Comparison Principle) *Assume that $u, v \colon S \to \mathbb{R}$ are solutions to the Dirichlet problem in Ω with boundary data α, β, respectively. We assume that*

$\alpha(x) \geq \beta(x)$ *for points* $x \in \partial\Omega$ *that are accessible by paths* $\gamma \in \mathcal{G}_0$, *where* $\mathcal{G}_0$ *is a path family that contains almost every path. Then* $u \geq v$ *in* S.

Proof Assume that the conclusion fails, so there exists $x_0 \in S$ with $u(x_0) < v(x_0)$. Let $f = u \vee v$ which is a Sobolev function with boundary values α on $\partial\Omega$. Thus, f is admissible for the Dirichlet problem on Ω with boundary values α, so $D_\Omega(u) \leq D_\Omega(f)$. By the uniqueness of solutions to the Dirichlet problem in Theorem 2.7.6, it follows that

$$D_\Omega(u) < D_\Omega(f). \tag{2.32}$$

Similarly, consider $g = u \wedge v$ which is admissible for the Dirichlet problem on Ω with boundary values β. As before, this implies that $D_\Omega(v) < D_\Omega(g)$. Adding this to (2.32), we obtain

$$\sum_{i\in\mathbb{N}} (\operatorname*{osc}_{Q_i}(u)^2 + \operatorname*{osc}_{Q_i}(v)^2) < \sum_{i\in\mathbb{N}} (\operatorname*{osc}_{Q_i}(f)^2 + \operatorname*{osc}_{Q_i}(g)^2).$$

This, however, contradicts (2.27) in Proposition 2.5.27.   □

2.7.3 Continuous Boundary Data

In this section we continue the treatment of the Dirichlet problem, proving that the solutions are continuous up to the boundary if the boundary data is continuous. Here we assume, as usual, that the boundary $\partial\Omega$ consists of finitely many disjoint Jordan curves.

Theorem 2.7.8 *Assume that* $u \colon S \to \mathbb{R}$ *is the solution to the Dirichlet problem in* Ω *with continuous boundary data* $f \colon \partial\Omega \to \mathbb{R}$. *Then* u *can be extended continuously to* $\partial\Omega$.

Proof The proof is very similar to the proof of Theorem 2.7.4 and uses, in some sense, a maximum principle near the boundary.

Recall that there exists a path family $\mathcal{G}_0$ that contains almost every path, such that for all $x \in \partial\Omega$ that are accessible by paths $\gamma \in \mathcal{G}_0$ we have $u(x) = f(x)$. Furthermore, the fact that the boundary $\partial\Omega$ consists of finitely many Jordan curves implies that every $x \in \partial\Omega$ is the landing point of a path $\gamma \subset \Omega$ (not necessarily in $\mathcal{G}_0$). Perturbing γ as in Lemma 2.4.4 we obtain a point $y \in \partial\Omega$ near x that is accessible by a path $\gamma_0 \in \mathcal{G}$. Hence, $u(y) = f(y)$ and this actually holds for a dense set of points in $\partial\Omega$.

We fix a point $x_0 \in \partial\Omega$ and we wish to show that u can extended at x_0 by $u(x_0) = f(x_0)$, so that $u\big|_{S\cup\{x_0\}}$ is continuous. If this is true for each $x_0 \in \partial\Omega$, then u will be continuous in $S \cup \partial\Omega$ by the continuity of f, as desired.

If $\mathrm{osc}_{Q_i}(u) = 0$ for all Q_i contained in a neighborhood of x_0, then by Lemma 2.7.2(b) u is a constant c near x_0. In particular, there exists an arc $\alpha \subset \partial\Omega$ with x_0 lying in the interior of α such that $u(y) = f(y) = c$ for a dense set of points $y \in \alpha$. This implies that $f(x_0) = c$ by continuity, and hence we may define $u(x_0) = c = f(x_0)$.

Now, suppose that arbitrarily close to x_0 we can find Q_i with $\mathrm{osc}_{Q_i}(u) > 0$. Consider a ball $B(x_0, r)$, where $r > 0$ is so small that $B(x_0, r)$ intersects only one boundary component of $\partial\Omega$. The boundary $\partial B(x_0, r)$ defines a crosscut $\gamma'_r \subset \partial B(x_0, r)$, which bounds a region $W \subset \Omega$, together with a subarc α of $\partial\Omega$, such that $x_0 \in \partial W$; see Fig. 2.2. We fix $\varepsilon > 0$ and take an even smaller r so that

$$\sup_{y \in \alpha} f(y) - \inf_{y \in \alpha} f(y) < \varepsilon \tag{2.33}$$

and $\sum_{i:Q_i \cap \gamma'_r \neq \emptyset} \mathrm{osc}_{Q_i}(u) < \varepsilon$, where the path $\gamma'_r \subset \gamma_r = \partial B(x_0, r)$ is non-exceptional, as in Lemma 2.4.7(a). We wish to show that

$$|f(x_0) - u(z)| \leq 2\varepsilon \tag{2.34}$$

for all $z \in S \cap W$. This will show that u can be continuously extended at x_0 by $u(x_0) = f(x_0)$.

Let $M = \sup_{y \in \alpha} f(y)$ and $m = \inf_{y \in \alpha} f(y)$. We claim that

$$m - \varepsilon \leq M_{Q_k}(u) \leq M + \varepsilon$$

for all $Q_k \subset W$. This will imply that $|f(x_0) - M_{Q_k}(u)| < 2\varepsilon$ by (2.33), and thus $|f(x_0) - u(z)| \leq 2\varepsilon$ for all $z \in S \cap W$, as desired; here we used the continuity of u and the fact that near z we can find arbitrarily small peripheral disks, and thus peripheral disks $Q_k \subset W$.

Observe that on the arc γ'_r by the upper gradient inequality we have $|u(x) - u(y)| \leq \sum_{i:Q_i \cap \gamma'_r} \mathrm{osc}_{Q_i}(u) < \varepsilon$. Since γ'_r is non-exceptional, we have $u(y) = f(y)$ for the endpoints of γ'_r. Hence

$$m - \varepsilon < u(x) \leq M + \varepsilon$$

for all $x \in S \cap \gamma'_r = \partial_* W$. Consider the function $h = u \cdot \chi_{S \setminus W} + u \wedge (M + \varepsilon) \cdot \chi_{S \cap W}$ which is a Sobolev function by Corollary 2.5.31 with $\mathrm{osc}_{Q_i}(h) \leq \mathrm{osc}_{Q_i}(u)$ for all $i \in \mathbb{N}$. Note that h is also admissible for the Dirichlet problem with boundary data f. If there exists some $Q_k \subset W$ with $M_{Q_k}(u) > M + \varepsilon$, then we can find actually some $Q_k \subset W$ with $M_{Q_k}(u) > M + \varepsilon$ and $\mathrm{osc}_{Q_k}(u) > 0$; see proof of Theorem 2.7.4. Then $\mathrm{osc}_{Q_k}(h) < \mathrm{osc}_{Q_k}(u)$ which contradicts the minimizing property of u. The inequality $m - \varepsilon \leq M_{Q_k}(u)$ for all $k \in \mathbb{N}$ is shown in the same way. □

So far we have treated the existence, the uniqueness, and the comparison principle for solutions to the Dirichlet problem. A natural question that arises is whether every carpet-harmonic function can be realized at least locally as a solution to a Dirichlet problem, so that we can apply these principles. It turns out that this is the case.

Proposition 2.7.9 *Let* $u: S \to \mathbb{R}$ *be a carpet-harmonic function, and* $V \subset\subset \Omega$ *be an open set with the properties:*

(1) ∂V *consists of finitely many disjoint Jordan curves, and*
(2) *if* $V \cap Q_i \neq \emptyset$ *then* $\overline{Q}_i \subset V$. *This, in particular, implies that* $\partial V \subset S$, *and* $(S \cap V, V)$ *is a relative Sierpiński carpet.*

Then u *agrees inside* $S \cap V$ *with the solution to the Dirichlet problem in* $S \cap V$ *with boundary values on* ∂V *equal to* u.

Proof Let $v: S \cap V \to \mathbb{R}$ be the solution to the Dirichlet problem with boundary values equal to u, given by Theorem 2.6.6. Since u is continuous, by Theorem 2.7.8 we have that v has a continuous extension to $S \cap \overline{V}$ that agrees with u on ∂V.

Consider the function $\zeta = (v - u) \cdot \chi_{S \cap V} + 0 \cdot \chi_{S \setminus V}$. This is a Sobolev function on S, but we cannot apply Lemma 2.5.28 directly, since v is not defined on all of S. The fact that ζ is a Sobolev function on S follows from the following lemma that we prove right after:

Lemma 2.7.10 *Let* $V \subset\subset \Omega$ *be an open set as above, with* $\partial V \subset S$ *and* $S \cap V$ *being a relative Sierpiński carpet. Consider functions* $\phi, \psi: S \to \mathbb{R}$ *such that* $\phi|_{S \cap V} \in \mathcal{W}_*^{1,2}(S \cap V)$, $\psi \in \mathcal{W}_{*,\mathrm{loc}}^{1,2}(S)$, *and* $\phi = \psi$ *on* ∂V. *Then* $\zeta := \phi \cdot \chi_{S \cap V} + \psi \cdot \chi_{S \setminus V}$ *lies in* $\mathcal{W}_{*,\mathrm{loc}}^{1,2}(S)$.

In our case we set $\psi \equiv 0$ and $\phi = (v - u) \cdot \chi_{S \cap V}$, which is continuous on S. The harmonicity of u implies that

$$D_V(u) \leq D_V(u + \zeta) = D_V(v).$$

However, v is a minimizer for the Dirichlet energy in V, and this implies that $D_V(u) = D_V(v)$. Using the uniqueness from Theorem 2.7.6, we conclude that $u = v$ on $S \cap V$. □

Proof of Lemma 2.7.10 Consider the families of good paths $\mathcal{G}_\phi, \mathcal{G}_\psi$ for ϕ, ψ, respectively. Note that the paths of $\mathcal{G}_\phi$ are contained in V. Let Γ_0 be the paths of $\mathcal{G}_\psi$ that have a subpath in V which is not contained in $\mathcal{G}_\phi$. Then one can show that Γ_0 has weak (strong) modulus equal to zero with respect to the carpet S; see also Remark 2.4.6. We define $\mathcal{G}$ to be the family of paths in $\mathcal{G}_\psi$ that do not lie in Γ_0, and we shall show ζ has an upper gradient and the upper gradient inequality holds along these paths.

Let $x, y \in \gamma \cap S$ and $\gamma \in \mathcal{G}$ be a path that connects them. If $x, y \notin V$, then

$$|\zeta(x) - \zeta(y)| = |\psi(x) - \psi(y)| \leq \sum_{i: Q_i \cap \gamma \neq \emptyset} \underset{Q_i}{\mathrm{osc}}(\psi).$$

If $x \in V$ and $y \notin V$, then $\gamma \cap \partial V \neq \emptyset$ and we can consider the point z of first entry of γ in ∂V, as it travels from x to y. The point z is accessible by γ, so by the definition of the boundary values of ϕ we have

$$\phi(z) = \liminf_{\substack{Q_i \to z \\ Q_i \cap \gamma \neq \emptyset, i \in I_V}} M_{Q_i}(\phi).$$

The upper gradient inequality therefore holds up to the boundary ∂V and we have

$$|\phi(x) - \phi(z)| \leq \sum_{\substack{i:Q_i \cap \gamma \neq \emptyset \\ i \in I_V}} \underset{Q_i}{\mathrm{osc}}(\phi).$$

Therefore,

$$|\zeta(x) - \zeta(y)| \leq |\phi(x) - \phi(z)| + |\psi(z) - \psi(y)|$$

$$\leq \sum_{i:Q_i \cap \gamma \neq \emptyset} (\underset{Q_i}{\mathrm{osc}}(\phi)\chi_{I_V}(i) + \underset{Q_i}{\mathrm{osc}}(\psi)).$$

The case $x, y \in V$ is treated similarly by considering also the point w of first entry of γ in ∂V, as it travels from y towards x, thus obtaining the bound

$$|\zeta(x) - \zeta(y)| \leq \sum_{i:Q_i \cap \gamma \neq \emptyset} (2\underset{Q_i}{\mathrm{osc}}(\phi)\chi_{I_V}(i) + \underset{Q_i}{\mathrm{osc}}(\psi)).$$

This shows that $\{2\,\mathrm{osc}_{Q_i}(\phi)\chi_{I_V}(i) + \mathrm{osc}_{Q_i}(\psi)\}_{i \in \mathbb{N}}$ is an upper gradient of ζ, as desired; recall Remark 2.5.13. $\qquad\square$

2.7.4 A Free Boundary Problem

In this section we mention some results on a different type of a boundary problem, in which boundary data is only present on part of the boundary. The proofs are almost identical to the case of the Dirichlet problem so we omit them. These results are used in Chap. 3 to prove a uniformization result for planar Sierpiński carpets.

Let $\Omega \subset \mathbb{C}$ be a quadrilateral, i.e., a Jordan region with four marked sides on $\partial\Omega$. Assume that the sides are closed and they are marked by $\Theta_1, \Theta_2, \Theta_3, \Theta_4$, in a counter-clockwise fashion, where Θ_1, Θ_3 are opposite sides. Consider a relative Sierpiński carpet (S, Ω). In fact, in this case $\overline{S}$ is an actual Sierpiński carpet, as defined in the Introduction, Chap. 1. We consider functions $g \in \mathcal{W}^{1,2}_*(S)$ with boundary data $g = 0$ on Θ_1 and $g = 1$ on Θ_3. Such functions are called *admissible* (*for the free boundary problem*).

Theorem 2.7.11 *There exists a unique carpet-harmonic function $u\colon S \to \mathbb{R}$ that minimizes the Dirichlet energy $D_\Omega(g)$ over all admissible functions $g \in \mathcal{W}^{1,2}_*(S)$. The function u is has a continuous extension to $\partial\Omega$ (and thus to $\overline{S}$), with $u = 0$ on Θ_1 and $u = 1$ on Θ_3.*

Of course, if the class of admissible functions is the weak Sobolev class then u is weak carpet-harmonic, and if the class of admissible functions is the strong Sobolev class, then u is strong carpet-harmonic.

Since there is no boundary data on the interior of the arcs Θ_2, Θ_4, these arcs can be treated—in the proofs—as subarcs of peripheral circles, and, in particular, as subsets of the carpet S. If $V \subset \mathbb{C} \setminus (\Theta_1 \cup \Theta_3)$ is an open set we can define

$$\partial_\bullet V = \partial V \cap \overline{S}.$$

This will be the boundary, on which the extremal values of the carpet-harmonic function u are attained. Thus, the maximum principle reads as:

Theorem 2.7.12 *If $u\colon \overline{S} \to \mathbb{R}$ is the solution to the free boundary problem, then for any open set $V \subset \mathbb{C} \setminus (\Theta_1 \cup \Theta_3)$ we have*

$$\sup_{x \in \overline{S} \cap \overline{V}} u(x) = \sup_{x \in \partial_\bullet V} u(x) \quad and \quad \inf_{x \in \overline{S} \cap \overline{V}} u(x) = \inf_{x \in \partial_\bullet V} u(x).$$

This is a stronger maximum principle than the one in Theorem 2.7.5. Note that for sets $V \subset\subset \Omega$ the two statements are the same. However, here we allow V to intersect Θ_2 and Θ_4, which are contained in $\partial\Omega$; see Fig. 2.5. Theorem 2.7.12 says that the extremal values of u on $\overline{S} \cap \overline{V}$ not only are attained in $\partial(V \cap \Omega) \cap S$, but they are actually attained at the part of $\partial(V \cap \Omega) \cap S$ that is disjoint from the interiors of the "free" arcs Θ_2 and Θ_4. However, this boundary could intersect Θ_1 and Θ_3, where extremal values can be attained, and this is the reason that we look at sets $V \subset \mathbb{C} \setminus (\Theta_1 \cup \Theta_3)$. See also the maximum principle as stated in [33, Lemma 4.6].

Fig. 2.5 An open set $V \subset \mathbb{C} \setminus (\Theta_1 \cup \Theta_3)$ (pink) intersecting a carpet $\overline{S}$, and the set $\partial_\bullet V = \partial V \cap \overline{S}$ that corresponds to V. Here, $\partial\Omega$ is the Jordan curve that separates S from ∞

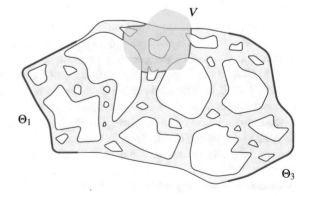

As an application, one can show a rigidity-type result for square Sierpiński carpets, which was established in [9, Theorem 1.4]:

Theorem 2.7.13 *Let* $\Omega = (0, 1) \times (0, A)$, $\Omega' = (0, 1) \times (0, A')$, *and consider relative Sierpiński carpets* $(S, \Omega), (S'\Omega')$ *such that all peripheral disks of* S, S' *are squares with sides parallel to the coordinate axes, and the Hausdorff 2-measure of* S *and* S' *is* 0. *If* $f : \overline{S} \to \overline{S'}$ *is a quasisymmetry that preserves the sides of* Ω, Ω' *(i.e.,* $f(\{0\} \times [0, A]) = \{0\} \times [0, A']$ *etc.) then* $A = A'$, $S = S'$, *and* f *is the identity.*

We remark that our proof here is simpler than the proof in [9], and relies on the uniqueness of Theorem 2.7.11; see also Theorem 2.7.6. In [9], the authors have to follow several steps, showing first that each square Q_i and its image Q'_i have the same side length, then that $Q_i = Q'_i$, and finally that f is the identity. They pursue these steps using the absolute continuity of f and modulus arguments. In our approach, these arguments seem to be incorporated in the properties of Sobolev spaces and in the uniqueness of the minimizer in Theorem 2.7.11, which is a powerful statement.

Proof Let u be the solution to the free boundary problem on (S, Ω) with $u = 0$ on $\Theta_1 := \{0\} \times [0, A]$ and $u = 1$ on $\Theta_3 := \{1\} \times [0, A]$. For $y \in [0, A]$ consider the segment $\gamma_y = [0, 1] \times \{y\}$ that is parallel to the x-axis. As in the proof of Lemma 2.4.3, one can see that for a.e. $y \in [0, A]$ the path γ_y is a good path for u, thus by the upper gradient inequality

$$1 = |u(0, y) - u(0, 1)| \leq \sum_{i : Q_i \cap \gamma_y \neq \emptyset} \underset{Q_i}{\mathrm{osc}}(u).$$

Integrating over $y \in [0, A]$ and applying the Cauchy–Schwarz inequality we obtain

$$A \leq \sum_{i \in \mathbb{N}} \underset{Q_i}{\mathrm{osc}}(u) \ell(Q_i) \leq \left(\sum_{i \in \mathbb{N}} \underset{Q_i}{\mathrm{osc}}(u)^2 \right)^{1/2} \left(\sum_{i \in \mathbb{N}} \ell(Q_i)^2 \right)^{1/2}.$$

The latter sum is the area of Ω, which is A, hence we obtain $A \leq D_\Omega(u)$. On the other hand, the function $g(x, y) = x$ is admissible for the free boundary problem, and it is easy to see that $A = D_\Omega(g)$. Since u is a minimizer it follows that $D_\Omega(g) = D_\Omega(u)$ and by the uniqueness in Theorem 2.7.11 we have $u(x, y) = g(x, y) = x$ for all $(x, y) \in \overline{S}$.

If $f : \overline{S} \to \overline{S'}$ is a quasisymmetry, then it extends to a quasiconformal homeomorphism $f = (u_0, v_0) : \overline{\Omega} \to \overline{\Omega'}$ (using $\mathbb{R}^2$ coordinates), by the extension results proved in [7, Sect. 5], and in fact, it extends to a global quasiconformal map on $\widehat{\mathbb{C}}$. Example 2.5.20 and Remark 2.5.22 show that f restricts to a function in $\mathcal{W}_s^{1,2}(S)$. Since f preserves the sides of Ω, it follows that $u_0 = 0$ on Θ_1 and $u_0 = 1$ on Θ_3. Hence, u_0 is admissible for the free boundary problem in Ω, and thus $A = D_\Omega(u) \leq D_\Omega(u_0)$. Note that $\mathrm{osc}_{Q_i}(u_0) = \ell(Q_i')$, where $Q_i' = f(Q_i)$. Hence

$$D_\Omega(u_0) = \sum_{i \in \mathbb{N}} \ell(Q_i')^2 = \mathcal{H}^2(\Omega') = A'.$$

It follows that

$$A = D_\Omega(u) \leq D_\Omega(u_0) = A'.$$

The same argument applied to f^{-1} and the free boundary problem in Ω' yields $A' \leq A$. Thus $D_\Omega(u) = D_\Omega(u_0) = A = A'$. The uniqueness in Theorem 2.7.11 shows that $u(x, y) = u_0(x, y) = x$.

The same argument applied to the dual free boundary problem $v = 0$ on $\Theta_2 := [0, 1] \times \{0\}$ and $v = A$ on $\Theta_4 := [0, 1] \times \{A\}$ yields $v_0(x, y) = y$. $\square$

2.8 The Caccioppoli Inequality

In this section, and also in the next, we assume that (S, Ω) is an arbitrary relative Sierpiński carpet (with the standard assumptions of Definition 2.2.1). We still drop the terminology weak/strong, and carpet-harmonic functions are always assumed to be normalized, as in Sect. 2.7. It will be convenient to use the term *test function* to refer to a function $\zeta \in \mathcal{W}_*^{1,2}(S)$ that vanishes outside an open set $V \subset\subset \Omega$.

Theorem 2.8.1 (Caccioppoli's Inequality) *Let* $\zeta : S \to \mathbb{R}$ *be a non-negative test function, and* $u : S \to \mathbb{R}$ *be a carpet-harmonic function. Then*

$$\sum_{i \in \mathbb{N}} M_{Q_i}(\zeta)^2 \mathop{\mathrm{osc}}_{Q_i}(u)^2 \leq C \sum_{i \in \mathbb{N}} \mathop{\mathrm{osc}}_{Q_i}(\zeta)^2 M_{Q_i}(|u|)^2,$$

where $C > 0$ *is some universal constant.*

The proof is an adaptation in the discrete carpet setting of the classical proof of Caccioppoli's inequality for harmonic functions.

Proof We can assume that ζ is bounded. Indeed, if ζ is unbounded, then for $M \in \mathbb{R}$ the function $\zeta \wedge M$ is a bounded Sobolev function. Moreover, we have $\mathrm{osc}_{Q_i}(\zeta \wedge M) \leq \mathrm{osc}_{Q_i}(\zeta)$ by Proposition 2.5.27(c) and $M_{Q_i}(\zeta \wedge M) \to M_{Q_i}(\zeta)$ as $M \to \infty$, which show that the desired inequality holds for ζ if it holds for $\zeta \wedge M$.

By assumption, $\zeta = 0$ outside a set $V \subset\subset \Omega$. For $\varepsilon > 0$ consider $\eta = \varepsilon\zeta^2$ and $h = u - \eta u$. The function η is a Sobolev function by Lemma 2.5.27(d), and so is ηu, by the same lemma and the local boundedness of the carpet-harmonic function u. Therefore, h is a Sobolev function that is equal to u outside V. It follows that $D_V(u) \leq D_V(h)$, by harmonicity. Now, we will estimate $\mathrm{osc}_{Q_i}(h)$.

We recall the computational rule from Lemma 2.5.27(d), which is similar to the product rule for derivatives: for all $i \in \mathbb{N}$ and all functions $f_1, f_2 \colon S \to \mathbb{R}$ we have

$$\operatorname*{osc}_{Q_i}(f_1 f_2) \le M_{Q_i}(|f_2|) \operatorname*{osc}_{Q_i}(f_1) + M_{Q_i}(|f_1|) \operatorname*{osc}_{Q_i}(f_2). \tag{2.35}$$

Using this rule, for fixed $i \in \mathbb{N}$ we have

$$\operatorname*{osc}_{Q_i}(h) = \operatorname*{osc}_{Q_i}(u(1-\eta)) \le M_{Q_i}(|1-\eta|) \operatorname*{osc}_{Q_i}(u) + M_{Q_i}(|u|) \operatorname*{osc}_{Q_i}(1-\eta).$$

Since ζ is bounded, for small $\varepsilon > 0$ we have $m_{Q_i}(\eta) \le M_{Q_i}(\eta) < 1$ for all $i \in \mathbb{N}$. This implies that $M_{Q_i}(|1-\eta|) = 1 - m_{Q_i}(\eta)$. Also, we trivially have $\operatorname{osc}_{Q_i}(1-\eta) = \operatorname{osc}_{Q_i}(\eta)$. Therefore, for all sufficiently small $\varepsilon > 0$ and for all $i \in \mathbb{N}$

$$\operatorname*{osc}_{Q_i}(h) \le (1 - m_{Q_i}(\eta)) \operatorname*{osc}_{Q_i}(u) + M_{Q_i}(|u|) \operatorname*{osc}_{Q_i}(\eta).$$

Combining this with the inequality $D_V(u) \le D_V(h)$ we obtain

$$\sum_{i \in I_V} \operatorname*{osc}_{Q_i}(u)^2 \le \sum_{i \in I_V} \Bigg[(1 - m_{Q_i}(\varepsilon\zeta^2))^2 \operatorname*{osc}_{Q_i}(u)^2 + M_{Q_i}(|u|)^2 \operatorname*{osc}_{Q_i}(\varepsilon\zeta^2)^2$$

$$+ 2(1 - m_{Q_i}(\varepsilon\zeta^2)) \operatorname*{osc}_{Q_i}(u) M_{Q_i}(|u|) \operatorname*{osc}_{Q_i}(\varepsilon\zeta^2) \Bigg].$$

Noting that $m_{Q_i}(\varepsilon\zeta^2) = \varepsilon m_{Q_i}(\zeta^2)$, $\operatorname{osc}_{Q_i}(\varepsilon\zeta^2) = \varepsilon \operatorname{osc}_{Q_i}(\zeta^2)$, and doing cancellations yields

$$0 \le \sum_{i \in I_V} \Bigg[(-2\varepsilon m_{Q_i}(\zeta^2) + \varepsilon^2 m_{Q_i}(\zeta^2)^2) \operatorname*{osc}_{Q_i}(u)^2 + \varepsilon^2 M_{Q_i}(|u|)^2 \operatorname*{osc}_{Q_i}(\zeta^2)^2$$

$$+ 2\varepsilon(1 - \varepsilon m_{Q_i}(\zeta^2)) \operatorname*{osc}_{Q_i}(u) M_{Q_i}(|u|) \operatorname*{osc}_{Q_i}(\zeta^2) \Bigg].$$

Dividing by $\varepsilon > 0$ and letting $\varepsilon \to 0$ we obtain

$$\sum_{i \in I_V} m_{Q_i}(\zeta^2) \operatorname*{osc}_{Q_i}(u)^2 \le \sum_{i \in I_V} \operatorname*{osc}_{Q_i}(u) M_{Q_i}(|u|) \operatorname*{osc}_{Q_i}(\zeta^2). \tag{2.36}$$

Now, we use the inequalities

$$\operatorname*{osc}_{Q_i}(\zeta^2) \le 2 M_{Q_i}(\zeta) \operatorname*{osc}_{Q_i}(\zeta) \quad \text{and}$$

$$m_{Q_i}(\zeta^2) = M_{Q_i}(\zeta^2) - \operatorname*{osc}_{Q_i}(\zeta^2) \ge M_{Q_i}(\zeta)^2 - 2 M_{Q_i}(\zeta) \operatorname*{osc}_{Q_i}(\zeta),$$

where the first one follows from the computational rule (2.35). Together with (2.36) they imply that

$$\sum_{i \in I_V} M_{Q_i}(\zeta)^2 \operatorname*{osc}_{Q_i}(u)^2 - 2 \sum_{i \in I_V} M_{Q_i}(\zeta) \operatorname*{osc}_{Q_i}(\zeta) \operatorname*{osc}_{Q_i}(u)^2$$

$$\leq 2 \sum_{i \in I_V} \operatorname*{osc}_{Q_i}(u) M_{Q_i}(|u|) M_{Q_i}(\zeta) \operatorname*{osc}_{Q_i}(\zeta).$$

In the second term of the left hand side we first use the inequality $\operatorname*{osc}_{Q_i}(u) \leq 2 M_{Q_i}(|u|)$, and then apply the Cauchy–Schwarz inequality:

$$\sum_{i \in I_V} M_{Q_i}(\zeta)^2 \operatorname*{osc}_{Q_i}(u)^2 \leq 6 \sum_{i \in I_V} \operatorname*{osc}_{Q_i}(u) M_{Q_i}(|u|) M_{Q_i}(\zeta) \operatorname*{osc}_{Q_i}(\zeta)$$

$$\leq 6 \left(\sum_{i \in I_V} M_{Q_i}(\zeta)^2 \operatorname*{osc}_{Q_i}(u)^2 \right)^{1/2} \cdot \left(\sum_{i \in I_V} \operatorname*{osc}_{Q_i}(\zeta)^2 M_{Q_i}(|u|)^2 \right)^{1/2}.$$

Hence

$$\sum_{i \in I_V} M_{Q_i}(\zeta)^2 \operatorname*{osc}_{Q_i}(u)^2 \leq 36 \sum_{i \in I_V} \operatorname*{osc}_{Q_i}(\zeta)^2 M_{Q_i}(|u|)^2.$$

Since $\zeta = 0$ outside V, we can in fact write the summations over $i \in \mathbb{N}$, and this completes the proof.   □

We now record an consequence of Caccioppoli's inequality. Namely, we prove a weak version of Liouville's theorem:

Theorem 2.8.2 *Let $(S, \mathbb{C})$ be a relative Sierpiński carpet, and $u \colon S \to \mathbb{R}$ a carpet-harmonic function such that $|u|$ is bounded. Then u is constant.*

The strong version of Liouville's theorem, where we merely assume that u is bounded above *or* below is proved in Sect. 2.9 using Harnack's inequality; see Corollary 2.9.3.

Proof Assume that $|u| \leq M$. We fix a ball $B(0, R_0)$ and we wish to construct a test function ζ such that $0 \leq \zeta \leq 1$, $\zeta = 1$ on $B(0, R_0)$, but $D_\mathbb{C}(\zeta)$ is arbitrarily small, not depending on R_0. Then by Caccioppoli's inequality we will have

$$\sum_{i \in I_{B(0,R_0)}} \operatorname*{osc}_{Q_i}(u)^2 \leq \sum_{i \in \mathbb{N}} M_{Q_i}(\zeta)^2 \operatorname*{osc}_{Q_i}(u)^2 \leq C \sum_{i \in \mathbb{N}} \operatorname*{osc}_{Q_i}(\zeta)^2 M_{Q_i}(|u|)^2 \leq CM^2 D_\mathbb{C}(\zeta).$$

Since $D_\mathbb{C}(\zeta)$ can be arbitrarily small, it follows that $\sum_{i : Q_i \cap B(0,R_0) \neq \emptyset} \operatorname*{osc}_{Q_i}(u)^2 = 0$, and thus $\operatorname*{osc}_{Q_i}(u) = 0$ for all Q_i that intersect $B(0, R_0)$. Since R_0 is arbitrary we have $\operatorname*{osc}_{Q_i}(u) = 0$ for all $i \in \mathbb{N}$. Thus, u is constant by Lemma 2.7.2(a), as desired.

Now, we construct the test function ζ with the desired properties. Roughly speaking, ζ will be a discrete version of the logarithm. We fix a large integer N which will correspond to the number of annuli around 0 that we will construct, and ζ will drop by $1/N$ on each annulus. Define $\zeta = 1$ on $B(0, R_0)$, and consider $r_1 := R_0$, $R_1 := 2r_1$. In the annulus $A_1 := A(0; r_1, R_1)$ define ζ to be the radial function of constant slope $\frac{1}{Nr_1}$, so that on the outer boundary of A_1 the function ζ has value $1 - 1/N$. Then consider $r_2 > R_1$ sufficiently large and $R_2 := 2r_2$, so that no peripheral disk intersects both A_1 and $A_2 := A(0; r_2, R_2)$; recall Lemma 2.3.4. In the transition annulus $A(0; R_1, r_2)$ we define ζ to be constant, equal to $1 - 1/N$, and on A_2 we let ζ be the radial function with slope $\frac{1}{Nr_2}$. The last annulus will be $A_N = A(0; r_N, R_N)$ and the value of ζ will be 0 on the outer boundary of A_N. We define ζ to be 0 outside $B(0, R_N)$. Note that ζ is locally Lipschitz, so its restriction to the carpet S is a Sobolev function, by Example 2.5.19.

We now compute the Dirichlet energy of ζ. Let $d_j(Q_i) = \mathcal{H}^1(\{s \in [r_j, R_j] : \gamma_s \cap Q_i \neq \emptyset\})$ where γ_s is the circle of radius s around 0. Since the peripheral disks Q_i are uniformly Ahlfors 2-regular, there exists a constant $K > 0$ such that $d_j(Q_i)^2 \leq K\mathcal{H}^2(Q_i \cap A_j)$; see Remark 2.3.5. Also, if $Q_i \cap A_j \neq \emptyset$, then $\operatorname*{osc}_{Q_i}(\zeta) \leq d_j(Q_i)\frac{1}{Nr_j}$. By construction, each peripheral disk Q_i can only intersect one annulus A_j, and if a peripheral disk Q_i does not intersect any annulus A_j, then ζ is constant on Q_i, so $\operatorname*{osc}_{Q_i}(\zeta) = 0$. Thus

$$
D_{\mathbb{C}}(\zeta) = \sum_{i \in \mathbb{N}} \operatorname*{osc}_{Q_i}(\zeta)^2 = \sum_{j=1}^{N} \sum_{i:Q_i \cap A_j \neq \emptyset} \operatorname*{osc}_{Q_i}(\zeta)^2
$$

$$
\leq \frac{1}{N^2} \sum_{j=1}^{N} \frac{1}{r_j^2} \sum_{i:Q_i \cap A_j \neq \emptyset} d_j(Q_i)^2
$$

$$
\leq \frac{K}{N^2} \sum_{j=1}^{N} \frac{1}{r_j^2} \sum_{i:Q_i \cap A_j \neq \emptyset} \mathcal{H}^2(Q_i \cap A_j)
$$

$$
\leq \frac{K}{N^2} \sum_{j=1}^{N} \frac{1}{r_j^2} \mathcal{H}^2(A_j)
$$

$$
= \frac{\pi K}{N^2} \sum_{j=1}^{N} \frac{R_j^2 - r_j^2}{r_j^2} = \frac{\pi K}{N^2} \sum_{j=1}^{N} \frac{3r_j^2}{r_j^2}
$$

$$
= \frac{3\pi K}{N},
$$

which can be made arbitrarily small if N is large. $\qquad\square$

Remark 2.8.3 Liouville's theorem (Theorem 2.8.2) justifies the choice to only define carpet-harmonic functions on relative carpets in planar domains, rather than in the whole sphere $\widehat{\mathbb{C}}$, i.e., $\Omega = \widehat{\mathbb{C}}$. Indeed, if we used $\Omega = \widehat{\mathbb{C}}$, then by continuity the carpet-harmonic functions would then be bounded and thus constant by Liouville's theorem. We point out that the setting here is different from Bonk's setting in [7], where the carpets are defined on the whole sphere $\widehat{\mathbb{C}}$.

The non-linearity of the theory does not allow us to apply Caccioppoli's inequality of Theorem 2.8.1 to linear combinations of harmonic functions. We record another version of the Caccioppoli inequality for differences of harmonic functions. This will be very useful in establishing convergence properties of harmonic functions in Sect. 2.10.

Theorem 2.8.4 *Let $\zeta : S \rightarrow \mathbb{R}$ be a non-negative continuous test function, and $u, v : S \rightarrow \mathbb{R}$ be carpet-harmonic functions. Then*

$$\sum_{i \in \mathbb{N}} m_{Q_i}(\zeta)(\underset{Q_i}{\operatorname{osc}}(u) - \underset{Q_i}{\operatorname{osc}}(v))^2 \leq C \sum_{i \in \mathbb{N}} \underset{Q_i}{\operatorname{osc}}(\zeta)(\underset{Q_i}{\operatorname{osc}}(u) + \underset{Q_i}{\operatorname{osc}}(v)) M_{Q_i}(|u - v|),$$

where $C > 0$ is some universal constant.

A similar inequality for harmonic functions in Carnot groups appears in [4] and in fact our proof is inspired by the proof of [4, Theorem 1.2].

Proof Suppose that $\eta : S \rightarrow \mathbb{R}$ is a continuous function and $\operatorname{osc}_{Q_i}(u + \varepsilon\eta) = u(x_\varepsilon) + \varepsilon\eta(x_\varepsilon) - u(y_\varepsilon) - \varepsilon\eta(y_\varepsilon)$ for some $\varepsilon \in \mathbb{R}$ and points $x_\varepsilon, y_\varepsilon \in \partial Q_i$. Then as $\varepsilon \rightarrow 0$, the points $x_\varepsilon, y_\varepsilon$ subconverge to points x, y, respectively, with $\operatorname{osc}_{Q_i}(u) = u(x) - u(y)$. Here we use the continuity of u and the boundedness of η on ∂Q_i.

If η is a continuous test function supported in $V \subset\subset \Omega$, we have $D_V(u) \leq D_V(u + \varepsilon\eta)$, which implies

$$\sum_{i \in I_V} \underset{Q_i}{\operatorname{osc}}(u)^2$$

$$\leq \sum_{i \in I_V} (u(x_{i,\varepsilon}) + \varepsilon\eta(x_{i,\varepsilon}) - u(y_{i,\varepsilon}) - \varepsilon\eta(y_{i,\varepsilon}))^2$$

$$\leq \sum_{i \in I_V} \underset{Q_i}{\operatorname{osc}}(u)^2 + 2\varepsilon \sum_{i \in I_V} (u(x_{i,\varepsilon}) - u(y_{i,\varepsilon}))(\eta(x_{i,\varepsilon}) - \eta(y_{i,\varepsilon})) + O(\varepsilon^2),$$

for some points $x_{i,\varepsilon}, y_{i,\varepsilon} \in \partial Q_i$. As $\varepsilon \rightarrow 0$, this yields

$$\sum_{i \in I_V} \underset{Q_i}{\operatorname{osc}}(u)(\eta(x_i) - \eta(y_i)) = 0, \tag{2.37}$$

where x_i, y_i are sublimits of $x_{i,\varepsilon}, y_{i,\varepsilon}$, respectively, and $\operatorname{osc}_{Q_i}(u) = u(x_i) - u(y_i)$.

We use $\eta = \zeta \cdot (u - v)$ in (2.37), where ζ is supported in $V \subset\subset \Omega$, and we obtain

$$0 = \sum_{i \in I_V} \operatorname*{osc}_{Q_i}(u)(\zeta(x_i)(u(x_i) - v(x_i)) - \zeta(y_i)(u(y_i) - v(y_i)))$$

$$= \sum_{i \in I_V} \operatorname*{osc}_{Q_i}(u) \cdot \bigg(\zeta(x_i)(u(x_i) - u(y_i)) - \zeta(x_i)(v(x_i) - v(y_i)) $$

$$+ (\zeta(x_i) - \zeta(y_i))(u(y_i) - v(y_i)) \bigg).$$

Since $\operatorname*{osc}_{Q_i}(u) = u(x_i) - u(y_i)$, $\operatorname*{osc}_{Q_i}(v) \geq v(x_i) - v(y_i)$, and $\zeta \geq 0$, we have

$$\sum_{i \in I_V} \zeta(x_i)(\operatorname*{osc}_{Q_i}(u)^2 - \operatorname*{osc}_{Q_i}(u) \operatorname*{osc}_{Q_i}(v)) \leq \sum_{i \in I_V} \operatorname*{osc}_{Q_i}(\zeta) \operatorname*{osc}_{Q_i}(u) M_{Q_i}(|u - v|).$$

Interchanging the roles of u and v, we obtain points $x_i' \in \partial Q_i$ such that

$$\sum_{i \in I_V} \zeta(x_i')(\operatorname*{osc}_{Q_i}(v)^2 - \operatorname*{osc}_{Q_i}(u) \operatorname*{osc}_{Q_i}(v)) \leq \sum_{i \in I_V} \operatorname*{osc}_{Q_i}(\zeta) \operatorname*{osc}_{Q_i}(v) M_{Q_i}(|u - v|).$$

Adding the two inequalities, we have

$$\sum_{i \in I_V} \zeta(x_i)(\operatorname*{osc}_{Q_i}(u) - \operatorname*{osc}_{Q_i}(v))^2 \leq \sum_{i \in I_V} \operatorname*{osc}_{Q_i}(\zeta)(\operatorname*{osc}_{Q_i}(u) + \operatorname*{osc}_{Q_i}(v)) M_{Q_i}(|u - v|)$$

$$+ \sum_{i \in I_V}(\zeta(x_i) - \zeta(x_i'))(\operatorname*{osc}_{Q_i}(v)^2 - \operatorname*{osc}_{Q_i}(u) \operatorname*{osc}_{Q_i}(v)).$$

Upon observing that $\zeta(x_i) \geq m_{Q_i}(\zeta)$, $|\zeta(x_i) - \zeta(x_i')| \leq \operatorname*{osc}_{Q_i}(\zeta)$, and

$$|\operatorname*{osc}_{Q_i}(u) - \operatorname*{osc}_{Q_i}(v)| \leq \operatorname*{osc}_{Q_i}(u - v) = M_{Q_i}(u - v) - m_{Q_i}(u - v) \leq 2M_{Q_i}(|u - v|),$$

we obtain the conclusion. $\qquad\qquad\square$

2.9 Harnack's Inequality and Consequences

2.9.1 Harnack's Inequality

In this section the main theorem is:

Theorem 2.9.1 (Harnack's Inequality) *There exists a constant $H > 1$ with the following property: if $u: S \to \mathbb{R}$ is a non-negative carpet-harmonic function, then*

$$\sup_{z \in S \cap B_0} u(z) \leq H \inf_{z \in S \cap B_0} u(z)$$

for all balls $B_0 \subset \Omega$ with the property that there exists a ball $B_1 \subset \Omega$ such that

$$\bigcup_{i:Q_i \cap c_0 B_0 \neq \emptyset} Q_i \subset B_1 \subset c_1 B_1 \subset\subset \Omega,$$

where $c_0, c_1 > 1$ are constants. The constant H depends only on the data of the carpet S and on c_0, c_1. The choice of the latter two constants can be arbitrary.

The assumption in the theorem asserts that the peripheral disks that meet $c_0 B_0$ are essentially "safely" contained in Ω, away from the boundary.

Our treatment of Harnack's inequality is inspired by [19], where Harnack's inequality is proved for $W^{1,n}$-minimizers of certain variational integrals in $\mathbb{R}^n$. The method used there is a purely variational argument, which does not rely on a differential equation or a representation formula for minimizers, and this allows us to apply it in our discrete setting. The proof will be done in several steps. First we show the following.

Proposition 2.9.2 *Let $u : S \to \mathbb{R}$ be a positive carpet-harmonic function. Then for any non-negative test function $\zeta : S \to \mathbb{R}$ we have*

$$\sum_{i \in \mathbb{N}} \frac{m_{Q_i}(\zeta)^2 \operatorname{osc}_{Q_i}(u)^2}{M_{Q_i}(u) m_{Q_i}(u)} \leq C D_\Omega(\zeta),$$

where $C > 0$ is a universal constant, not depending on u, ζ, S.

Note that by continuity u is bounded below away from 0 on each individual peripheral circle ∂Q_i, and this shows that $M_{Q_i}(u) \geq m_{Q_i}(u) > 0$. Hence, all quantities above make sense.

Proof As in the proof of Theorem 2.8.1, we may assume that ζ is bounded. By replacing u with $u + \delta$ for a small $\delta > 0$, and noting that $\operatorname{osc}_{Q_i}(u + \delta) = \operatorname{osc}_{Q_i}(u)$, $M_{Q_i}(u + \delta) = M_{Q_i}(u) + \delta$, $m_{Q_i}(u + \delta) = m_{Q_i}(u) + \delta$, we see that it suffices to prove the statement assuming that $u \geq \delta > 0$. The constant $C > 0$ will of course be independent of δ, so that we can let $\delta \to 0$.

Fix a bounded test function ζ, supported in $S \cap V$, where $V \subset\subset \Omega$. For a small $\varepsilon > 0$ consider the variation $h = u + \varepsilon \zeta^2 / u$. We remark that the function $u^{-1} = 1/u$ is a Sobolev function, since

$$|u(x)^{-1} - u(y)^{-1}| \leq |u(x) - u(y)|/\delta^2,$$

so u^{-1} inherits its upper gradient inequality from u. Moreover, ζ^2 and $\zeta^2 u^{-1}$ are Sobolev functions by Lemma 2.5.27(d). Hence, h is a Sobolev function.

Observe that the function $a \mapsto a + \varepsilon b^2/a$ is increasing as long as $a^2 > \varepsilon b^2$. Since ζ is bounded and $u \geq \delta$, it follows that for sufficiently small $\varepsilon > 0$ we have $u^2 \geq \delta^2 \geq \varepsilon(\sup_{x \in S} \zeta(x))^2$. Thus for all $i \in \mathbb{N}$ and $x \in \partial Q_i$ we have

$$m_{Q_i}(u) + \varepsilon \frac{\zeta(x)^2}{m_{Q_i}(u)} \leq h(x) \leq M_{Q_i}(u) + \varepsilon \frac{\zeta(x)^2}{M_{Q_i}(u)}.$$

This implies that for all sufficiently small $\varepsilon > 0$ we have

$$M_{Q_i}(h) \leq M_{Q_i}(u) + \varepsilon \frac{M_{Q_i}(\zeta^2)}{M_{Q_i}(u)}, \quad \text{and}$$

$$m_{Q_i}(h) \geq m_{Q_i}(u) + \varepsilon \frac{m_{Q_i}(\zeta^2)}{m_{Q_i}(u)}$$

for all $i \in \mathbb{N}$. Hence,

$$\operatorname*{osc}_{Q_i}(h) \leq \operatorname*{osc}_{Q_i}(u) + \varepsilon \left(\frac{M_{Q_i}(\zeta^2)}{M_{Q_i}(u)} - \frac{m_{Q_i}(\zeta^2)}{m_{Q_i}(u)} \right)$$

$$= \left(1 - \varepsilon \frac{m_{Q_i}(\zeta^2)}{M_{Q_i}(u) m_{Q_i}(u)} \right) \operatorname*{osc}_{Q_i}(u) + \varepsilon \frac{M_{Q_i}(\zeta) + m_{Q_i}(\zeta)}{M_{Q_i}(u)} \operatorname*{osc}_{Q_i}(\zeta),$$

where we used the equalities $\operatorname*{osc}_{Q_i}(\zeta) = M_{Q_i}(\zeta) - m_{Q_i}(\zeta)$ and $\operatorname*{osc}_{Q_i}(\zeta^2) = \operatorname*{osc}_{Q_i}(\zeta) \cdot (M_{Q_i}(\zeta) + m_{Q_i}(\zeta))$.

Since u is carpet-harmonic, and h is equal to u outside V we have $D_V(u) \leq D_V(h)$. Working as in the proof of Theorem 2.8.1, and letting $\varepsilon \to 0$, we obtain

$$\sum_{i \in I_V} \frac{m_{Q_i}(\zeta^2) \operatorname*{osc}_{Q_i}(u)^2}{M_{Q_i}(u) m_{Q_i}(u)} \leq \sum_{i \in I_V} \operatorname*{osc}_{Q_i}(u) \frac{M_{Q_i}(\zeta) + m_{Q_i}(\zeta)}{M_{Q_i}(u)} \operatorname*{osc}_{Q_i}(\zeta).$$

Writing $M_{Q_i}(\zeta) = m_{Q_i}(\zeta) + \operatorname*{osc}_{Q_i}(\zeta)$ and applying the Cauchy–Schwarz inequality we obtain

$$\sum_{i \in I_V} \frac{m_{Q_i}(\zeta^2) \operatorname*{osc}_{Q_i}(u)^2}{M_{Q_i}(u) m_{Q_i}(u)} \leq \sum_{i \in I_V} \frac{\operatorname*{osc}_{Q_i}(u)}{M_{Q_i}(u)} (2 m_{Q_i}(\zeta) + \operatorname*{osc}_{Q_i}(\zeta)) \operatorname*{osc}_{Q_i}(\zeta)$$

$$\leq 2 \left(\sum_{i \in I_V} \frac{m_{Q_i}(\zeta)^2 \operatorname*{osc}_{Q_i}(u)^2}{M_{Q_i}(u)^2} \right)^{1/2} D_V(\zeta)^{1/2}$$

$$+ \sum_{i \in I_V} \frac{\operatorname*{osc}_{Q_i}(u)}{M_{Q_i}(u)} \operatorname*{osc}_{Q_i}(\zeta)^2.$$

We define $A = \sum_{i \in I_V} \frac{m_{Q_i}(\zeta)^2 \operatorname{osc}_{Q_i}(u)^2}{M_{Q_i}(u) m_{Q_i}(u)}$. Noting that $M_{Q_i}(u)^2 \geq M_{Q_i}(u) m_{Q_i}(u)$ and $\operatorname{osc}_{Q_i}(u) = M_{Q_i}(u) - m_{Q_i}(u) \leq M_{Q_i}(u)$, we have

$$A \leq 2A^{1/2} D_V(\zeta)^{1/2} + D_V(\zeta).$$

If $A \leq D_V(\zeta)$, then there is nothing to show. Otherwise, we have $D_V(\zeta)^{1/2} \leq A^{1/2}$, hence

$$A \leq 2A^{1/2} D_V(\zeta)^{1/2} + A^{1/2} D_V(\zeta)^{1/2},$$

which implies that

$$A \leq 9 D_V(\zeta)$$

and this concludes the proof.   □

This proposition already has the strong Liouville theorem as a corollary:

Corollary 2.9.3 (Liouville's Theorem) *Let $(S, \mathbb{C})$ be a relative Sierpiński carpet, and $u \colon S \to \mathbb{R}$ be a carpet-harmonic function that is bounded above or below. Then u is constant.*

Proof We first reduce the statement to the case that $u > 0$. If u is bounded above, then we can replace it with the carpet-harmonic function $1 + \sup_{z \in S} u(z) - u$, and showing that this is constant will imply that u is constant. Similarly, if u is bounded below, then we use $1 + u - \inf_{z \in S} u(z)$. Hence, from now on we assume that $u > 0$.

We fix a ball $B(0, R_0)$ and consider a test function ζ such that $0 \leq \zeta \leq 1$, $\zeta = 1$ on $S \cap B(0, R_0)$, and $D_{\mathbb{C}}(\zeta) < \varepsilon$ where $\varepsilon > 0$ can be arbitrarily small. Such a function is constructed in the proof of Theorem 2.8.2. Then for all $Q_i \subset B(0, R_0)$ we have $m_{Q_i}(\zeta) = 1$. Hence, Proposition 2.9.2 yields

$$\sum_{i : Q_i \subset B(0, R_0)} \frac{\operatorname{osc}_{Q_i}(u)^2}{M_{Q_i}(u) m_{Q_i}(u)} \leq C D_{\mathbb{C}}(\zeta) < C\varepsilon.$$

Letting $\varepsilon \to 0$ we obtain $\operatorname{osc}_{Q_i}(u) = 0$ for all $Q_i \subset B(0, R_0)$. Since R_0 was arbitrary, it follows that $\operatorname{osc}_{Q_i}(u) = 0$ for all $i \in \mathbb{N}$, thus u is constant by Lemma 2.7.2(a).   □

We continue our preparation for the proof of Harnack's inequality. From Proposition 2.9.2 we derive the next lemma:

Lemma 2.9.4 *Let $u \colon S \to \mathbb{R}$ be a positive carpet-harmonic function. Consider a ball $B_1 \subset c_1 B_1 \subset\subset \Omega$ for some $c_1 > 1$. Then*

$$\sum_{i : Q_i \subset B_1} \frac{\operatorname{osc}_{Q_i}(u)^2}{M_{Q_i}(u) m_{Q_i}(u)} \leq C,$$

where the constant $C > 0$ depends only on the data of the carpet S and on c_1, but not on u, B_1.

Proof We apply Proposition 2.9.2 to a test function ζ defined as follows. We set $\zeta = 1$ on B_1, $\zeta = 0$ outside $c_1 B_1$, and ζ is radial with slope $\frac{1}{(c_1-1)r}$ on the annulus $A_1 = c_1 B_1 \setminus \overline{B}_1$, where r is the radius of B_1. Then ζ is Lipschitz so by Example 2.5.19 it restricts to a Sobolev function. We only have to show that $D_\Omega(\zeta)$ is bounded by a constant depending only on c_1 and the data of S. Our computation is very similar to the proof of Theorem 2.8.2. We introduce the notation $d(Q_i) = \mathcal{H}^1(\{s \in [r, c_1 r] : \gamma_s \cap Q_i \neq \emptyset\})$ and note that $d(Q_i)^2 \leq K\mathcal{H}^2(Q_i \cap c_1 B_1)$ for $i \in \mathbb{N}$ by the Ahlfors regularity condition and Remark 2.3.5. Then

$$D_\Omega(\zeta) = \sum_{i \in I_{A_1}} \operatorname*{osc}_{Q_i}(\zeta)^2 \leq \frac{1}{(c_1-1)^2 r^2} \sum_{i \in I_{A_1}} d(Q_i)^2$$

$$\leq \frac{K}{(c_1-1)^2 r^2} \mathcal{H}^2(c_1 B_1) = \frac{K c_1^2 \pi}{(c_1-1)^2},$$

as claimed. Note that the latter expression diverges to ∞ as $c_1 \to 1$.   □

Next we prove a version of Gehring's oscillation lemma (see e.g. [2, Lemma 3.5.1, p. 65]). A function $v \colon S \to \mathbb{R}$ is said to be *monotone* if it satisfies the maximum and minimum principles as in the statement of Theorem 2.7.5.

Lemma 2.9.5 *Let $v \colon S \to \mathbb{R}$ be a continuous monotone function on the relative Sierpiński carpet (S, Ω), lying in the Sobolev space $\mathcal{W}^{1,2}_{*,\mathrm{loc}}(S)$. Consider a ball $B_0 \subset \Omega$ with $B_0 \subset c_0 B_0 \subset\subset \Omega$, where $c_0 > 1$. Then*

$$\sup_{z \in S \cap B_0} v(z) - \inf_{z \in S \cap B_0} v(z) \leq C \left(\sum_{i \in I_{c_0 B_0}} \operatorname*{osc}_{Q_i}(v)^2 \right)^{1/2},$$

where the constant $C > 0$ depends only on the data of the carpet S and on c_0, but not on v, B_0.

Proof For any $x, y \in S \cap B_0$, by monotonicity we have

$$v(x) - v(y) \leq \sup_{z \in \partial_*(s B_0)} v(z) - \inf_{z \in \partial_*(s B_0)} v(z) \tag{2.38}$$

for all $s \in [1, c_0]$. For a.e. $s \in [1, c_0]$ the upper gradient inequality for v yields

$$\sup_{z \in \partial_*(s B_0)} v(z) - \inf_{z \in \partial_*(s B_0)} v(z) \leq \sum_{i : Q_i \cap \partial(s B_0) \neq \emptyset} \operatorname*{osc}_{Q_i}(v). \tag{2.39}$$

Here we used the fact that the circular path $\partial(sB_0)$ is non-exceptional for a.e. $s \in [1, c_0]$, which follows from the proof of Lemma 2.4.3. If we write $B_0 = B(x_0, r)$, then (2.38) and (2.39) imply that

$$\sup_{z \in S \cap B_0} v(z) - \inf_{z \in S \cap B_0} v(z) \leq \sum_{i : Q_i \cap \partial B(x_0, s) \neq \emptyset} \operatorname*{osc}_{Q_i}(v)$$

for a.e. $s \in [r, c_0 r]$. We now integrate over $s \in [r, c_0 r]$ so we obtain

$$r(c_0 - 1)\left(\sup_{z \in S \cap B_0} v(z) - \inf_{z \in S \cap B_0} v(z)\right) \leq \int_r^{c_0 r} \sum_{i : Q_i \cap \partial B(x_0, s) \neq \emptyset} \operatorname*{osc}_{Q_i}(v)\, ds$$

$$= \sum_{i \in \mathbb{N}} \operatorname*{osc}_{Q_i}(v) \int_r^{c_0 r} \chi_{Q_i \cap \partial B(x_0, s)}\, ds$$

$$\leq \sum_{i : Q_i \cap c_0 B_0 \neq \emptyset} \operatorname*{osc}_{Q_i}(v) d(Q_i),$$

where $d(Q_i) = \mathcal{H}^1(\{s \in [r, c_0 r] : \gamma_s \cap Q_i \neq \emptyset\})$ and γ_s is a circular path around x_0 with radius s. As usual, by the Ahlfors regularity of the peripheral disks (see Remark 2.3.5) there exists a uniform constant K such that $d(Q_i)^2 \leq K\mathcal{H}^2(Q_i \cap c_0 B_0)$ for all $i \in \mathbb{N}$. Now, applying Cauchy–Schwarz we obtain

$$\sup_{z \in S \cap B_0} v(z) - \inf_{z \in S \cap B_0} v(z) \leq \frac{1}{r(c_0 - 1)}\left(\sum_{i \in I_{c_0 B_0}} \operatorname*{osc}_{Q_i}(v)^2\right)^{1/2}\left(\sum_{i \in I_{c_0 B_0}} d(Q_i)^2\right)^{1/2}$$

$$\leq \frac{1}{r(c_0 - 1)}\left(\sum_{i \in I_{c_0 B_0}} \operatorname*{osc}_{Q_i}(v)^2\right)^{1/2} (K\mathcal{H}^2(c_0 B_0))^{1/2}$$

$$\leq \frac{K^{1/2}\pi^{1/2}c_0}{c_0 - 1}\left(\sum_{i \in I_{c_0 B_0}} \operatorname*{osc}_{Q_i}(v)^2\right)^{1/2}.$$

This completes the proof.   □

Finally we proceed with the proof of Harnack's inequality.

Proof of Theorem 2.9.1 Replacing u with $u + \delta$ for a small $\delta > 0$ and noting that the conclusion of the theorem persists if we let $\delta \to 0$, we may assume that $u \geq \delta > 0$.

The function $v = \log u$ is a continuous monotone function, since u has these properties and $\log$ is an increasing function. Also, v lies in the Sobolev space $\mathcal{W}^{1,2}_{*,\mathrm{loc}}(S)$, since

$$|v| \le \max\{|u|, |\log \delta|\}, \quad \text{and}$$

$$|v(x) - v(y)| = \left|\log \frac{u(x)}{u(y)}\right| \le |u(x) - u(y)|/\delta.$$

The latter inequality shows that v inherits its upper gradient inequality from u. In fact, we have

$$\operatorname*{osc}_{Q_i}(v) = \operatorname*{osc}_{Q_i}(\log u) \le \frac{\operatorname{osc}_{Q_i}(u)}{M_{Q_i}(u)^{1/2} m_{Q_i}(u)^{1/2}}, \tag{2.40}$$

as one can see from the elementary inequality $\log(a/b) \le \frac{a-b}{(ab)^{1/2}}$ for $a \ge b > 0$.

Now, applying Lemma 2.9.5, and then (2.40) and Lemma 2.9.4 one has

$$\sup_{z \in S \cap B_0} v(z) - \inf_{z \in S \cap B_0} v(z) \le C \left(\sum_{i \in I_{c_0 B_0}} \operatorname*{osc}_{Q_i}(v)^2 \right)^{1/2}$$

$$\le C \left(\sum_{i: Q_i \subset B_1} \operatorname*{osc}_{Q_i}(v)^2 \right)^{1/2}$$

$$\le C \left(\sum_{i: Q_i \subset B_1} \frac{\operatorname{osc}_{Q_i}(u)^2}{M_{Q_i}(u) m_{Q_i}(u)} \right)^{1/2}$$

$$\le C'.$$

Here we used the assumption that

$$\bigcup_{i \in I_{c_0 B_0}} Q_i \subset B_1 \subset c_1 B_1 \subset\subset \Omega.$$

Therefore

$$\log\left(\frac{\sup_{z \in S \cap B_0} u(z)}{\inf_{z \in S \cap B_0} u(z)}\right) = \sup_{z \in S \cap B_0} v(z) - \inf_{z \in S \cap B_0} v(z) \le C',$$

thus

$$\sup_{z \in S \cap B_0} u(z) \le e^{C'} \inf_{z \in S \cap B_0} u(z).$$

The constant $e^{C'}$ depends only on the data of the carpet S and on c_0, c_1. □

We record an application of the oscillation Lemma 2.9.5.

Corollary 2.9.6 *Let $(S, \mathbb{C})$ be a relative Sierpiński carpet, and $u: S \to \mathbb{R}$ a carpet-harmonic function with finite energy, i.e., $D_{\mathbb{C}}(u) < \infty$. Then u is constant.*

Proof By Lemma 2.9.5, for any ball $B_0 \subset \mathbb{C}$ we have

$$\sup_{z \in S \cap B_0} u(z) - \inf_{z \in S \cap B_0} u(z) \leq C \left(\sum_{i \in I_{2B_0}} \underset{Q_i}{\mathrm{osc}}(u)^2 \right)^{1/2} \leq C D_{\mathbb{C}}(u)^{1/2}.$$

The ball B_0 is arbitrary, so it follows that u is bounded, and therefore it is constant by Liouville's Theorem 2.9.3. $\square$

2.9.2 Strong Maximum Principle

Using Harnack's inequality we prove a strong maximum principle:

Theorem 2.9.7 *Let $u: S \to \mathbb{R}$ be a carpet-harmonic function. Assume that u attains a maximum or a minimum at a point $x_0 \in S$. Then u is constant.*

Proof Using $-u$ instead of u if necessary, we assume that x_0 is a point of maximum. First assume that $x_0 \in S^\circ$, i.e., x_0 does not lie on any peripheral circle. Let $v = u(x_0) - u$ which is a non-negative carpet-harmonic function. Then using Lemma 2.3.4 one can find small balls B_0, B_1 centered at x_0 such that

$$\bigcup_{i: Q_i \cap 2B_0 \neq \emptyset} Q_i \subset B_1 \subset 2B_1 \subset\subset \Omega.$$

Applying Harnack's inequality inside B_0 we obtain

$$\sup_{z \in S \cap B_0} v(z) \leq C \min_{z \in S \cap B_0} v(z) = 0.$$

Thus, $u(z) = u(x_0)$ for $z \in S \cap B_0$.

Now, if $y_0 \in S^\circ$ is arbitrary, then by Lemma 2.4.10 one can find a path $\gamma \subset S^\circ$ that connects y_0 to x_0. For each point $y \in \gamma$ there exists a small ball $B_y \subset \Omega$ where Harnack's inequality can be applied. By compactness, there are finitely many balls B_{y_i}, $i = 1, \ldots, N$, that cover the path γ and form a *Harnack chain*: B_{y_i} intersects $B_{y_{i+1}}$ and Harnack's inequality can be applied to each ball B_{y_i}. The argument in the previous paragraph yields that u is constant on each ball, thus $u(y_0) = u(x_0)$. Since S° is dense in S, by continuity it follows that u is constant.

Now, we treat the case that $x_0 \in \partial Q_{i_0}$ for some $i_0 \in \mathbb{N}$. Then by Lemma 2.4.10 there exists Jordan curve $\gamma \subset S^\circ$ enclosing a Jordan region V containing Q_{i_0}. By the maximum principle (Theorem 2.7.5), there exists a point $x_0' \in \gamma = \partial_* V$ such that $u(x_0') = \sup_{z \in S \cap \overline{V}} u(z) = u(x_0)$. Since $x_0' \in S^\circ$, it follows that u is constant by the previous case. $\square$

2.10   Equicontinuity and Convergence

Finally, we establish the local equicontinuity of carpet-harmonic functions and ensure that limits of harmonic functions are harmonic. In all statements the underlying relative Sierpiński carpet is (S, Ω).

Theorem 2.10.1 *Let $V \subset\subset U \subset\subset \Omega$ be open sets and $M > 0$ be a constant. For each $\varepsilon > 0$ there exists $\delta > 0$ such that if u is a carpet-harmonic function with $D_U(u) \leq M$, then for all points $x, y \in S \cap V$ with $|x - y| < \delta$ we have $|u(x) - u(y)| < \varepsilon$. The value of δ depends only on the carpet S and on ε, M, but not on u.*

Usually, equicontinuity for minimizers in potential theory follows from the local Hölder continuity of the energy minimizers. However, in our setting we were not able to establish the Hölder continuity, mainly due to the lack of self-similarity of the carpets.

Proof By compactness, it suffices to show that for each $\varepsilon > 0$, each point $x_0 \in S \cap \overline{V}$ has a neighborhood V_0 such that for all $z, w \in S \cap V_0$ we have $|u(z) - u(w)| < \varepsilon$, whenever u is a carpet-harmonic function with $D_U(u) \leq M$. The proof is based on the arguments we used in Theorem 2.8.2 and Lemma 2.9.5.

Suppose first that $x_0 \in S^\circ$, and let $N \in \mathbb{N}$ be a large number to be chosen. For each $k \in \{1, \ldots, N\}$ consider an annulus $A_k = A(x_0; r_k, 2r_k) \subset\subset U$ such that the annuli are nested, all of them surrounding $B(x_0, r_N)$, and they intersect disjoint sets of peripheral disks. Let $V_0 = B(x_0, r_N)$ and note that for $z, w \in S \cap V_0$ and $k \in \{1, \ldots, N\}$ we have

$$|u(z) - u(w)|^2 \leq C \sum_{i \in I_{A_k}} \operatorname*{osc}_{Q_i}(u)^2$$

by the computations in the proof of Lemma 2.9.5, where $C > 0$ depends only on the data. Summing over k, we obtain

$$N|u(z) - u(w)|^2 \leq C \sum_{i \in I_U} \operatorname*{osc}_{Q_i}(u)^2 = C D_U(u) \leq CM.$$

Hence, $|u(z) - u(w)| \leq C'\sqrt{M}/\sqrt{N}$, which can be made smaller than ε, if N is sufficiently large, independent of u.

If $x_0 \in \partial Q_{i_0}$ for some $i_0 \in \mathbb{N}$, we have to modify the argument as usual. We consider again the annuli A_k, all of which intersect Q_{i_0}, but otherwise they intersect disjoint sets of peripheral disks. We set V_0 to be the component of $B(x_0, r_N) \setminus$

$\overline{Q}_{i_0}$ containing x_0 in its boundary. This component is bounded by a subarc γ of $\partial B(x_0, r_N)$, which defines a crosscut separating x_0 from ∞ in $\mathbb{R}^2 \setminus Q_{i_0}$. The arc γ has its endpoints on ∂Q_{i_0}. We consider an open Jordan arc $\alpha \subset Q_{i_0}$, having the same endpoints as γ. Then $\gamma \cup \alpha$ bounds a Jordan region $V_1 \supset V_0$.

For $z, w \in S \cap V_1$ the maximum principle in Theorem 2.7.5 and the upper gradient inequality yield

$$|u(z) - u(w)| \leq \sum_{\substack{i: Q_i \cap \partial B(x_0, sr_k) \neq \emptyset \\ i \neq i_0}} \operatorname*{osc}_{Q_i}(u)$$

for all $k \in \{1, \ldots, N\}$ and a.e. $s \in (1, 2)$. Using this in the proof of Lemma 2.9.5 we obtain the exact same inequality as in the conclusion, without the term corresponding to $i = i_0$. Now, the proof continues as in the case $x_0 \in S$.   □

Theorem 2.10.2 *Suppose that u_n, $n \in \mathbb{N}$, is a sequence of carpet-harmonic functions converging locally uniformly to a function $u : S \to \mathbb{R}$. Then u is carpet-harmonic.*

Proof We fix an open set $V \subset\subset \Omega$. Then u_n converges uniformly to u in V, so in particular, u_n is uniformly bounded in V. By the Caccioppoli inequality in Theorem 2.8.1 we obtain that $D_V(u_n)$ is uniformly bounded in $n \in \mathbb{N}$. The Caccioppoli inequality for differences in Theorem 2.8.4 implies that $\sum_{i \in I_V}(\operatorname*{osc}_{Q_i}(u_n) - \operatorname*{osc}_{Q_i}(u_m))^2$ is uniformly small for sufficiently large n and m. This shows that the tails of the sum

$$D_V(u_n) = \sum_{i \in I_V} \operatorname*{osc}_{Q_i}(u_n)^2$$

are small, uniformly in n. Using the uniform convergence one can show that $\operatorname*{osc}_{Q_i}(u_n) \to \operatorname*{osc}_{Q_i}(u)$ for each $i \in \mathbb{N}$. These imply that $\{\operatorname*{osc}_{Q_i}(u_n)\}_{i \in V}$ converges to $\{\operatorname*{osc}_{Q_i}(u)\}_{i \in I_V}$ in ℓ^2. Hence, u lies in $\mathcal{W}^{1,2}_{*,\mathrm{loc}}(S)$, as follows from Remark 2.5.23.

Moreover, $D_V(u_n + \zeta) \to D_V(u + \zeta)$ for each test function $\zeta \in \mathcal{W}^{1,2}_*(S)$ vanishing outside V. Indeed, it is straightforward from uniform convergence to see that $\operatorname*{osc}_{Q_i}(u_n + \zeta) \to \operatorname*{osc}_{Q_i}(u + \zeta)$ for each $i \in \mathbb{N}$. Moreover, using the inequality $\operatorname*{osc}_{Q_i}(u_n + \zeta) \leq \operatorname*{osc}_{Q_i}(u_n) + \operatorname*{osc}_{Q_i}(\zeta)$ (see the proof of Proposition 2.5.27) and the convergence of $\{\operatorname*{osc}_{Q_i}(u_n)\}_{i \in V}$ in ℓ^2 one sees that the tails of the sum

$$\sum_{i \in I_V} \operatorname*{osc}_{Q_i}(u_n + \zeta)^2$$

are small, uniformly in n.

Finally, the above imply that the inequality $D_V(u_n) \leq D_V(u_n + \zeta)$ from harmonicity passes to the limit, to yield $D_V(u) \leq D_V(u + \zeta)$. This shows that u is carpet-harmonic, as desired.   □

Corollary 2.10.3 *Let u_n, $n \in \mathbb{N}$, be a sequence of carpet-harmonic functions that are locally uniformly bounded. Then there exists a subsequence of u_n that converges locally uniformly to a carpet-harmonic function $u : S \to \mathbb{R}$.*

Proof By the Caccioppoli inequality in Theorem 2.8.1, it follows that $D_V(u_n)$ is uniformly bounded in n, for each $V \subset\subset \Omega$. Theorem 2.10.1 implies that $\{u_n\}_{n \in \mathbb{N}}$ is a locally equicontinuous family. Since the functions u_n are locally uniformly bounded, by the Arzelà–Ascoli theorem we conclude that they subconverge locally uniformly to a function $u : S \to \mathbb{R}$. This function has to be carpet-harmonic by Theorem 2.10.2. $\square$

Chapter 3
Uniformization of Sierpiński Carpets by Square Carpets

3.1 Introduction

In this chapter we prove a uniformization result for planar Sierpiński carpets by square Sierpiński carpets, by minimizing some kind of energy. For the convenience of the reader, we include here some definitions, some of which are also given in Chap. 2. We will also point out, whenever necessary, any discrepancies in the notation between the two chapters. However, for the most part, this chapter is independent of Chap. 2 and we will only use certain results from there that we quote again here.

Before proceeding to the results, we mention some important discrepancies in the notation between the two chapters that the reader should be aware of:

1. A Sierpiński carpet is denoted here by $S = \overline{\Omega} \setminus \bigcup_{i \in \mathbb{N}} Q_i$, in contrast to Chap. 2, where the letter S was used to denote a relative Sierpiński carpet $S = \Omega \setminus \bigcup_{i \in \mathbb{N}} Q_i$.
2. ∂_* has slightly different meaning; compare its definition in Sect. 2.2 to the remarks after Theorem 3.4.5.
3. *Carpet modulus* is a variant of the strong carpet modulus defined in Sect. 2.3. Namely, the only difference is that here we also include the unbounded peripheral disk in the sums, whenever we have a path family in $\mathbb{C}$ and not necessarily in Ω. Otherwise, the definition is the same as in Chap. 2.
4. We will define a notion of modulus called *weak carpet modulus* that is slightly different from the weak carpet modulus defined in Sect. 2.3. One difference is that we again include the unbounded peripheral disk in the sums. The other difference is that a non-negative weight λ is admissible for a curve family Γ if

$$\sum_{i : \overline{Q}_i \cap \gamma \neq \emptyset} \lambda(Q_i) \geq 1$$

© The Editor(s) (if applicable) and The Author(s), under exclusive licence
to Springer Nature Switzerland AG 2020
D. Ntalampekos, *Potential Theory on Sierpiński Carpets*, Lecture Notes
in Mathematics 2268, https://doi.org/10.1007/978-3-030-50805-0_3

for all curves $\gamma \in \Gamma$ outside a curve family of 2-modulus equal to zero; see Definition 3.9.1. In contrast, for the weak carpet modulus of Chap. 2 we required instead that $\sum_{i:Q_i \cap \gamma \neq \emptyset} \lambda(Q_i) \geq 1$.

3.1.1 Results

In what follows, all the distances are in the Euclidean metric of $\mathbb{C}$.

A planar Sierpiński carpet $S \subset \mathbb{C}$ is constructed by removing from a Jordan region $\Omega \subset \mathbb{C}$ a countable collection $\{Q_i\}_{i \in \mathbb{N}}$ of open Jordan regions, compactly contained in Ω, with disjoint closures, such that $\operatorname{diam}(Q_i) \to 0$ as $i \to \infty$ and such that $S = \overline{\Omega} \setminus \bigcup_{i \in \mathbb{N}} Q_i$ has empty interior. The condition $\operatorname{diam}(Q_i) \to 0$ is equivalent to saying that S is locally connected. According to a fundamental result of Whyburn [44] all Sierpiński carpets are homeomorphic to each other. We remark that S is a closed set here, in contrast to Chap. 2, where $S \subset \Omega$; if we used the notation of Chap. 2, our carpet here would correspond to $\overline{S}$.

The Jordan regions Q_i, $i \in \mathbb{N}$, are called the *inner peripheral disks* of the carpet S, and $Q_0 := \mathbb{C} \setminus \overline{\Omega}$ is the *outer peripheral disk*. The Jordan curves ∂Q_i, $i \in \mathbb{N} \cup \{0\}$, are the *peripheral circles* of the carpet. Again, we distinguish between the *inner* peripheral circles ∂Q_i, $i \in \mathbb{N}$, and the outer peripheral circle ∂Q_0. A *square Sierpiński carpet* is a Sierpiński carpet for which Ω is a rectangle and all peripheral disks Q_i, $i \in \mathbb{N}$, are squares such that the sides of ∂Q_i, $i \in \mathbb{N} \cup \{0\}$, are parallel to the coordinate axes of $\mathbb{R}^2$.

We say that the peripheral disks Q_i are *uniformly quasiround*, if there exists a constant $K_0 \geq 1$ such that for each Q_i, $i \in \mathbb{N}$, there exist concentric balls $B(x, r)$, $B(x, R)$ with the property that

$$B(x, r) \subset Q_i \subset B(x, R), \tag{3.1}$$

and $R/r \leq K_0$. In this case, we also say that the peripheral disks are K_0-*quasiround*. We say that the peripheral disks are *uniformly Ahlfors 2-regular* if there exists a constant $K_1 > 0$ such that for every Q_i, $i \in \mathbb{N}$, and for every ball $B(x, r)$ centered at some $x \in Q_i$ with $r < \operatorname{diam}(Q_i)$ we have

$$\mathcal{H}^2(B(x, r) \cap Q_i) \geq K_1 r^2, \tag{3.2}$$

where by $\mathcal{H}^2$ we denote the two-dimensional Hausdorff measure, normalized so that it agrees with the two-dimensional Lebesgue measure. In this case, we say that the peripheral disks are K_1-*Ahlfors 2-regular sets*. A Jordan curve $J \subset \mathbb{C}$ is a K_2-*quasicircle* for some $K_2 > 0$, if for any two points $x, y \in J$ there exists an arc $\gamma \subset J$ with endpoints x, y such that

$$|x - y| \leq K_2 \operatorname{diam}(\gamma). \tag{3.3}$$

Note that if the peripheral circles ∂Q_i are uniform quasicircles (i.e., K_2-quasicircles with the same constant K_2), then they are both uniformly quasiround and uniformly Ahlfors 2-regular sets, quantitatively. A proof of the first claim can be found [7, Proposition 4.3] and the second claim is proved in [37, Corollary 2.3], where the notion of an Ahlfors 2-regular set appeared for the first time in the study of conformal maps. It is clear that for a square carpet the inner peripheral circles, also called peripheral squares, are uniform quasicircles.

In order to describe our main result we need to introduce a notion of quasiconformality suitable for the carpet setting. For this purpose, we introduce *carpet modulus with respect to the carpet S*. Let Γ be a family of paths in $\mathbb{C}$. A sequence of non-negative numbers $\{\lambda(Q_i)\}_{i\in\mathbb{N}\cup\{0\}}$ is admissible for the *carpet modulus* $\mathrm{mod}(\Gamma)$ if

$$\sum_{i: Q_i \cap \gamma \neq \emptyset} \lambda(Q_i) \geq 1 \tag{3.4}$$

for all $\gamma \in \Gamma$ with $\mathcal{H}^1(\gamma \cap S) = 0$. Then $\mathrm{mod}(\Gamma) := \inf_\lambda \sum_{i\in\mathbb{N}\cup\{0\}} \lambda(Q_i)^2$ where the infimum is taken over all admissible weights λ. Because of technical difficulties we also consider a very similar notion of modulus denoted by $\overline{\mathrm{mod}}(\Gamma)$, which is called *weak carpet modulus*; see Definition 3.9.1. The notation $\mathrm{mod}(\Gamma)$ and $\overline{\mathrm{mod}}(\Gamma)$ does not incorporate the underlying carpet S, but this will be implicitly understood. We now state one of the main results, which is illustrated in Fig. 3.1.

Theorem 3.1.1 *Let* $S \subset \overline{\Omega}$ *be a planar Sierpiński carpet of area zero whose peripheral disks* $\{Q_i\}_{i\in\mathbb{N}}$ *are uniformly quasiround and uniformly Ahlfors 2-regular, and whose outer peripheral circle, denoted by* ∂Q_0, *is* $\partial\Omega$. *Then there exists* $D > 0$ *and a homeomorphism* $f: \mathbb{C} \to \mathbb{C}$ *such that* $f(\overline{\Omega}) = [0, 1] \times [0, D]$, *and* $\mathcal{R} := f(S) \subset [0, 1] \times [0, D]$ *is a square Sierpiński carpet with inner peripheral squares* $\{S_i\}_{i\in\mathbb{N}}$ *and outer peripheral circle* $\partial([0, 1] \times [0, D])$, *denoted by* ∂S_0. *Furthermore, for any disjoint, non-trivial continua* $E, F \subset S$ *and for the family* Γ *of paths in* $\mathbb{C}$ *that join them we have*

$$\overline{\mathrm{mod}}(\Gamma) \leq \mathrm{mod}(f(\Gamma)) \quad \textit{and} \quad \overline{\mathrm{mod}}(f(\Gamma)) \leq \mathrm{mod}(\Gamma).$$

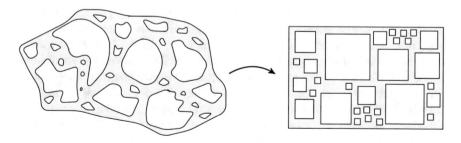

Fig. 3.1 Illustration of the uniformizing map in Theorem 3.1.1

A homeomorphism f between Sierpiński carpets satisfying the above inequalities is called *carpet-quasiconformal*. This definition is motivated by the so-called *geometric definition* of quasiconformality, which is also employed in [33]. Specifically, a map $f: \mathbb{C} \to \mathbb{C}$ is K-*quasiconformal* for some $K \geq 1$, according to the geometric definition, if for any curve family Γ in $\mathbb{C}$ we have

$$\frac{1}{K} \operatorname{mod}_2(\Gamma) \leq \operatorname{mod}_2(f(\Gamma)) \leq K \operatorname{mod}_2(\Gamma).$$

In this case we say that f *quasi-preserves modulus*. For quasiconformal maps in the plane as above this definition agrees with the so-called *analytic definition* of quasiconformality and in fact, quasiconformal maps of the plane are *quasisymmetric*; see Sect. 3.3.1 for the definitions. In other spaces though, such as carpets, a map that quasi-preserves modulus is not quasisymmetric in general. Hence, the notion of quasisymmetry is, in general, different from the notion of quasiconformality according to the geometric definition. As we see below in Proposition 3.1.6, carpet-quasiconformal maps are not expected to be quasisymmetric without any further geometric assumptions on the carpets.

Moreover, if we impose some stronger geometric assumptions, as in Theorem 3.1.2, then indeed the carpet-quasiconformal map f of Theorem 3.1.1 is quasisymmetric. This result follows closely the philosophy of [33], where it is first proved that a metric 2-sphere can be mapped quasiconformally onto to the Euclidean 2-sphere under some geometric assumptions. If the geometric assumptions are strengthened, it is proved then that a quasiconformal map from a metric 2-sphere onto the Euclidean 2-sphere enjoys further regularity and is actually quasisymmetric.

The map f in Theorem 3.1.1 is highly non-unique, as will be clear from the construction, since by making some choices one can obtain such maps onto different square carpets. We also remark that f is not quasiconformal in the usual sense (as a map from the plane to itself). Indeed, if one alters the map f arbitrarily inside a peripheral disk, then the carpet modulus inequalities above are not affected, so the resulting map is still carpet-quasiconformal.

We provide some brief justification for the mixture of carpet modulus and weak carpet modulus in the conclusion of Theorem 3.1.1. If the map $f|_S$ had a quasiconformal extension to the whole plane $\mathbb{C}$, then it would preserve the weak carpet modulus. The preservation of weak carpet modulus under quasiconformal maps of $\mathbb{C}$ was observed by Bonk and Merenkov [9, Lemma 2.1]. The reason is that the definition of weak carpet modulus (Definition 3.9.1) involves the conformal 2-modulus, which is quasi-preserved under quasiconformal maps. However, under our weak geometric assumptions, $f|_S$ does not have a quasiconformal extension to $\mathbb{C}$ and it is not clear if there is any other general type of extension that preserves 2-modulus. Because of the lack of an extension, we need to use another type of modulus that has an intrinsic definition (with respect to the carpet) and whose properties do not depend on some extension result. This is the carpet modulus that appears in the right hand sides of the conclusion of Theorem 3.1.1. The carpet modulus and weak carpet modulus are known to be equal to each other in some very special cases (see e.g. the first step of the proof of Theorem 3.11.1) but their relation is not clear in general. For example, they are not equal to each other when Γ is the family of curves passing through a point $p \in \mathbb{C} \setminus S$; see Remark 2.3.7. See Sect. 2.3 for more connections between the two types of carpet modulus.

As we have already discussed, if the geometric assumptions on the peripheral disks Q_i of S are strengthened, then we obtain a stronger version of the previous theorem with improved regularity for the map f. In particular, we consider the following geometric condition: we say that the peripheral circles $\{\partial Q_i\}_{i \in \mathbb{N} \cup \{0\}}$ of the carpet S are *uniformly relatively separated*, if there exists a constant $K_3 > 0$ such that

$$\Delta(\partial Q_i, \partial Q_j) := \frac{\operatorname{dist}(\partial Q_i, \partial Q_j)}{\min\{\operatorname{diam}(\partial Q_i), \operatorname{diam}(\partial Q_j)\}} \geq K_3 \qquad (3.5)$$

for all distinct $i, j \in \mathbb{N} \cup \{0\}$. In this case, we say that the peripheral circles are K_3-*relatively separated*. We now have an improvement of the previous theorem:

Theorem 3.1.2 *Let S be a Sierpiński carpet of area zero with peripheral circles $\{\partial Q_i\}_{i \in \mathbb{N} \cup \{0\}}$ that are K_2-quasicircles and K_3-relatively separated. Then there exists an η-quasisymmetric map f from S onto a square Sierpiński carpet $\mathcal{R}$ such that the distortion function η depends only on K_2 and K_3. Furthermore, the map f maps the outer peripheral circle of S onto the outer peripheral circle of $\mathcal{R}$.*

This theorem, together with extension results from [7, Sect. 5] yield immediately:

Corollary 3.1.3 *Let S be a Sierpiński carpet of area zero with peripheral circles $\{\partial Q_i\}_{i \in \mathbb{N} \cup \{0\}}$ that are K_2-quasicircles and K_3-relatively separated. Then there exists a K-quasiconformal map $f : \widehat{\mathbb{C}} \to \widehat{\mathbb{C}}$ such that $\mathcal{R} := f(S)$ is a square Sierpiński carpet, where K depends only on K_2 and K_3.*

Theorem 3.1.2 and its corollary should be compared to the result of Bonk:

Theorem 3.1.4 ([7, Corollary 1.2]) *Let S be a Sierpiński carpet of area zero whose peripheral circles $\{\partial Q_i\}_{i \in \mathbb{N} \cup \{0\}}$ are K_2-quasicircles and they are K_3-relatively separated. Then S can be mapped to a round Sierpiński carpet $T \subset \mathbb{C}$ with peripheral circles $\{\partial C_i\}_{i \in \mathbb{N} \cup \{0\}}$ by an η-quasisymmetric homeomorphism $f : S \to T$ that maps the outer peripheral circle ∂Q_0 of S to the outer peripheral circle ∂C_0 of T. The distortion function η depends only on K_2 and K_3, and the map f is unique up to post-composition with Möbius transformations.*

Here a *round Sierpiński carpet* is a Sierpiński carpet all of whose peripheral circles are round circles, i.e., circles in the usual geometric sense. Round carpets are in some sense natural objects, since they are invariant under Möbius transformations of the sphere and they are related to other subjects of Geometric Function theory, such as circle domains and Koebe's conjecture, Schottky groups and their limit sets etc. On the other hand, the square Sierpiński carpets studied here are also natural. First of all, the standard Sierpiński carpet (see Fig. 1.1 in Chap. 1) is a square carpet and any other carpet is homeomorphic to the standard carpet. Furthermore, square carpets appear naturally as "extremal" objects for a specific minimization problem that we consider here. Namely, the map f in Theorem 3.1.1 and in Theorem 3.1.2 is constructed by solving first a minimization problem, which gives rise to the real part u of f; see Sect. 3.4.

In our theorem, as already remarked, we do not have a strong uniqueness statement as in the result of Bonk in Theorem 3.1.4. Nevertheless, we can obtain a uniqueness result.

Proposition 3.1.5 *Assume that f is an orientation-preserving quasisymmetry from a Sierpiński carpet S of measure zero onto a square Sierpiński carpet $\mathcal{R}$ that maps the outer peripheral circle of S to the outer peripheral circle of $\mathcal{R}$, which is a rectangle $\partial([0, 1] \times [0, D])$. Let $\Theta_1 = f^{-1}(\{0\} \times [0, D])$ and $\Theta_3 = f^{-1}(\{1\} \times [0, D])$ be the preimages of the two vertical sides of $\partial([0, 1] \times [0, D])$. Assume also that g is another orientation-preserving quasisymmetry from S onto some other square carpet $\mathcal{R}'$, whose outer peripheral circle is a rectangle $\partial([0, 1] \times [0, D'])$. If g maps Θ_1 onto $\{0\} \times [0, D']$ and Θ_3 onto $\{1\} \times [0, D']$, then $g = f$ and $\mathcal{R}' = \mathcal{R}$.*

This proposition follows from [9, Theorem 1.4], or from Theorem 2.7.13 in Chap. 2 using the theory of carpet-harmonic functions.

Partial motivation for the current work was the desire to understand necessary conditions for the geometry of the peripheral disks of a carpet in order to obtain a quasisymmetric uniformization result. If one aims to obtain a uniformization result by round or square carpets, then a necessary assumption is that one of uniform quasicircles, since this quality is preserved under quasisymmetries, and both circles and squares share it. On the other hand, a round or square Sierpiński carpet need not satisfy the uniform relative separation condition, so the role of this condition is less clear. However, the next result implies that the condition is optimal in the case of mapping round carpets onto square carpets, and vice versa.

Proposition 3.1.6 *A square Sierpiński carpet $\mathcal{R}$ of area zero is quasisymmetrically equivalent to a round Sierpiński carpet T if and only if its peripheral circles are uniformly relatively separated. Conversely, a round Sierpiński carpet T of area zero is quasisymmetrically equivalent to a square Sierpiński carpet $\mathcal{R}$ if and only if its peripheral circles are uniformly relatively separated.*

Here, $S \subset \mathbb{C}$ is *quasisymmetrically equivalent* to $T \subset \mathbb{C}$ if there exists a quasisymmetry $f \colon S \to T$. Note that a round caret T of area zero satisfies the geometric assumptions of the main Theorem 3.1.1. Hence, there exists a carpet-quasiconformal map from T onto a square carpet $\mathcal{R}$. However, Proposition 3.1.6 implies that such a carpet-quasiconformal map is not quasisymmetric if the peripheral circles of T are not uniformly relatively separated.

Bonk's uniformizing map in Theorem 3.1.4 is constructed as a limit of conformal maps of finitely connected domains onto finitely connected circle domains, using Koebe's uniformization theorem. The first finitely connected domains converge in some sense to the carpet S and the latter finitely connected circle domains converge to a round carpet. Then, modulus estimates are used to study the properties of the limiting map, and show that it is a quasisymmetry.

In our approach, we construct an "extremal" map from the carpet to a square carpet by working directly on the carpet, without employing a limiting argument based on finitely connected domains. This has the advantage that we can control the necessary assumptions on the peripheral circles at each step of the construction. Although we impose the assumptions of uniformly Ahlfors 2-regular, uniformly quasiround sets for the peripheral disks $\{Q_i\}_{i \in \mathbb{N}}$, one can actually obtain Theorem 3.1.1 even if no assumptions at all are imposed on finitely many peripheral circles. To simplify the treatment we do not pursue this here, but we support our claim by remarking that no assumptions are imposed on the outer peripheral circle ∂Q_0, in the statement of Theorem 3.1.1.

3.1.2 Organization of the Chapter

In Sect. 3.2 we introduce some notation, and in Sect. 3.3 we discuss preliminaries on quasisymmetric maps, quasiconformal maps, modulus, and exceptional families of paths. The latter will be used extensively throughout the chapter.

The proof of the Main Theorem 3.1.1 will be given in Sects. 3.4–3.9. First, in Sect. 3.4 we introduce the real part u of the uniformizing map f. To obtain the map u we use the theory of carpet-harmonic functions, developed in Chap. 2. In fact, u will be a solution to a certain boundary value problem. In Sect. 3.5 we study the geometry of the level sets of the function u. The maximum principle is used in combination with an upper gradient inequality to deduce that almost every level set of u is the intersection of a curve with the carpet S, and in fact this intersection has Hausdorff 1-measure zero. The latter is the most technical result of the section.

By "integrating" the "gradient" of u along each level set, we define the conjugate function v of u in Sect. 3.6. The proof of continuity and regularity properties of v occupies the section. In Sect. 3.7 we define the uniformizing function $f = (u, v)$ that maps the carpet S into a rectangle $[0, 1] \times [0, D]$ and the peripheral circles into squares. We prove that the squares are disjoint and they occupy a full measure subset of the rectangle $[0, 1] \times [0, D]$. In Sect. 3.8 the injectivity of f is established and one of the main lemmas in the section is to show that f cannot "squeeze" a continuum $E \subset S$ to a single point. Finally, in Sect. 3.9 we prove regularity properties for f and f^{-1} and, in particular, the properties claimed in the Main Theorem 3.1.1.

Theorem 3.1.2 is proved in Sect. 3.11 with the aid of Loewner-type estimates for carpet modulus that are quoted in Sect. 3.10 and were proved (in some other form) by Bonk in [7]. Proposition 3.1.6 is proved in Sect. 3.12. Finally, in Sect. 3.13 we construct a *test function* that will be used frequently in variational arguments against the carpet-harmonic function u.

3.2   Notation and Terminology

We use the notation $\widehat{\mathbb{R}} = \mathbb{R} \cup \{-\infty, +\infty\}$, $\widehat{\mathbb{C}} = \mathbb{C} \cup \{\infty\}$. A function that attains values in $\widehat{\mathbb{R}}$ is called an *extended function*. We use the standard open ball notation $B(x, r) = \{y \in \mathbb{C} : |x - y| < r\}$ and denote a closed ball by $\overline{B}(x, r)$. If $B = B(x, r)$ then cB denotes the ball $B(x, cr)$. Also, $A(x; r, R)$ denotes the annulus $B(x, R) \setminus \overline{B}(x, r)$, for $0 < r < R$. All the distances will be measured using the Euclidean distance of $\mathbb{C} \simeq \mathbb{R}^2$. The variable x will denote most of the times a point of $\mathbb{R}^2$ and on rare occasion we will use the notation (x, y) for coordinates of a point in $\mathbb{R}^2$, in which case $x, y \in \mathbb{R}$. Each case will be clear from the context.

For $\alpha > 0$ the α-*dimensional Hausdorff measure* $\mathcal{H}^\alpha(E)$ of a set $E \subset \mathbb{C}$ is defined by

$$\mathcal{H}^\alpha(E) = \lim_{\delta \to 0} \mathcal{H}^\alpha_\delta(E) = \sup_{\delta > 0} \mathcal{H}^\alpha_\delta(E),$$

where

$$\mathcal{H}^\alpha_\delta(E) := \inf \left\{ \sum_{j=1}^\infty \operatorname{diam}(U_j)^\alpha : E \subset \bigcup_j U_j, \operatorname{diam}(U_j) < \delta \right\}.$$

The quantity $\mathcal{H}^\alpha_\delta(E)$ is called the α-*dimensional Hausdorff δ-content* of E. For $\alpha = 2$, we may normalize $\mathcal{H}^2$ by multiplying it with a constant, so that it agrees with the two-dimensional Lebesgue measure. We will use this normalization throughout the chapter.

The notation $V \subset\subset \Omega$ means that $\overline{V}$ is compact and is contained in Ω. For a set $K \subset \mathbb{C}$ and $\varepsilon > 0$ we use the notation

$$N_\varepsilon(K) = \{x \in \mathbb{C} : \operatorname{dist}(x, K) < \varepsilon\}$$

for the open ε-neighborhood of K. The symbols $\overline{V}$, $\operatorname{int}(V)$, and ∂V denote the closure, interior and boundary, respectively, of a set V with respect to the planar topology. If the reference space is a different set U, then we will indicate this for example by writing "*the closure of V rel. U*", or use subscript notation $\operatorname{int}_U(V)$.

A *path* or *curve* γ is a continuous function $\gamma : I \to \mathbb{C}$, where I is a bounded interval such that γ has a continuous extension $\overline{\gamma} : \overline{I} \to \mathbb{C}$, i.e., γ has endpoints. An *open path* γ is a path with $I = (0, 1)$. We will also use the notation $\gamma \subset \mathbb{C}$ for the image of the path as a set. A *subpath* of a path $\gamma : I \to \mathbb{C}$ is the restriction of γ to a subinterval of I. A path γ *joins* two sets $E, F \subset \mathbb{C}$ if $\overline{\gamma} \cap E \neq \emptyset$ and $\overline{\gamma} \cap F \neq \emptyset$. More generally, a connected set $\alpha \subset \mathbb{C}$ *joins* two sets $E, F \subset \mathbb{C}$ if $\overline{\alpha} \cap E \neq \emptyset$ and $\overline{\alpha} \cap F \neq \emptyset$. A Jordan curve is a homeomorphic image of the unit circle S^1, and a Jordan arc is a homeomorphic image of $[0, 1]$. Jordan curves and Jordan arcs are *simple* curves, i.e., they have no self-intersections.

We denote by S° the points of the Sierpiński carpet S that do not lie on any peripheral circle ∂Q_i or on $\partial \Omega$. For a set V that intersects a Sierpiński carpet S with (inner) peripheral disks $\{Q_i\}_{i \in \mathbb{N}}$ define $I_V = \{i \in \mathbb{N} : Q_i \cap V \neq \emptyset\}$.

In the proofs we will denote constants by $C, C', C'', \ldots$, where the same symbol might denote a different constant if there is no ambiguity. For visual purposes, the closure of a set U_1 is denoted by $\overline{U}_1$, instead of $\overline{U_1}$.

3.3 Preliminaries

3.3.1 Quasisymmetric and Quasiconformal Maps

A map $f \colon X \to Y$ between two metric spaces (X, d_X) and (Y, d_Y) is a *quasisymmetry* if it is a homeomorphism and there exists an increasing homeomorphism $\eta \colon (0, \infty) \to (0, \infty)$ such that for all triples of distinct points $x, y, z \in X$ we have

$$\frac{d_Y(f(x), f(y))}{d_Y(f(x), f(z))} \leq \eta \left(\frac{d_X(x, y)}{d_X(x, z)} \right).$$

The homeomorphism η is also called the distortion function associated to the quasisymmetry f and we say in this case that f is an η-*quasisymmetry*. If $X = Y = \mathbb{C}$, then it is immediate to see that the quasisymmetric property of $f \colon \mathbb{C} \to \mathbb{C}$ and the distortion function η are not affected by compositions with Möbius transformations of $\mathbb{C}$ (e.g., homotheties). A Jordan curve $J \subset \mathbb{C}$ is a quasicircle in the sense of (3.3) if and only if there exists a quasisymmetry $f \colon S^1 \to J$. The quasicircle constant and the distortion function of f are related quantitatively. See [21, Chaps. 10–11] for background on quasisymmetric maps.

Let $U, V \subset \mathbb{C}$ be open sets. An orientation-preserving homeomorphism $f \colon U \to V$ is K-*quasiconformal* for some $K \geq 1$ if $f \in W^{1,2}_{\text{loc}}(U)$ and

$$\|Df(z)\|^2 \leq K J_f(z)$$

for a.e. $z \in U$. An orientation-preserving homeomorphism $f \colon \widehat{\mathbb{C}} \to \widehat{\mathbb{C}}$ is K-quasiconformal if f is K-quasiconformal in local coordinates as a map between planar open sets, using the standard conformal charts of $\widehat{\mathbb{C}}$. We direct the reader to [2, Sect. 3] for more background.

3.3.2 Modulus

We recall the definition of *conformal modulus* or *2-modulus*. Let Γ be a path family in $\mathbb{C}$. A non-negative Borel extended function $\lambda: \mathbb{C} \to [0, \infty]$ is *admissible* for the conformal modulus $\mathrm{mod}_2(\Gamma)$ if

$$\int_\gamma \lambda \, ds \geq 1$$

for all locally rectifiable paths $\gamma \in \Gamma$. If a path γ is not locally rectifiable, we define $\int_\gamma \lambda \, ds = \infty$, even when $\lambda \equiv 0$. Hence, we may require the above inequality for all $\gamma \in \Gamma$. Then $\mathrm{mod}_2(\Gamma) := \inf_\lambda \int \lambda^2 \, d\mathcal{H}^2$, where the infimum is taken over all admissible functions λ.

Let us mention a connection between conformal modulus and the carpet modulus defined in the Introduction of the current chapter (Sect. 3.1). Let $S \subset \overline{\Omega}$ be a carpet with inner peripheral disks $\{Q_i\}_{i \in \mathbb{N}}$ and outer peripheral disk $Q_0 = \mathbb{C} \setminus \overline{\Omega}$. Suppose that $\mathcal{H}^2(S) = 0$ and the peripheral disks are uniformly Ahlfors 2-regular, uniformly quasiround. Consider a family Γ of paths contained in $\mathbb{C}$.

Lemma 3.3.1 *If* $\mathrm{mod}(\Gamma) = 0$ *then* $\mathrm{mod}_2(\Gamma) = 0$.

For the proof see Lemma 2.3.1 and Lemma 2.3.3.

If a property (A) holds for all paths γ in $\mathbb{C}$ lying outside an exceptional family of 2-modulus zero, we say that (A) *holds for* mod_2-*a.e.* γ. Furthermore, if the property (A) holds for mod_2-a.e. path in $\mathbb{C}$, then mod_2-a.e. path has the property that all of its subpaths also share property (A). Equivalently, the family of paths having a subpath for which property (A) fails has 2-modulus zero. The reason is that the family of admissible functions for this curve family contains the admissible functions for the family of curves for which property (A) fails; see also [43, Theorem 6.4]. More generally, let Γ be a path family such that, for each $\gamma \in \Gamma$, all of the subpaths of γ also lie in Γ (e.g., this holds if Γ is the family of all paths in $\mathbb{C}$ as above). If the property (A) holds for mod_2-a.e. path in Γ, then mod_2-a.e. path in Γ has the property that all of its subpaths also share property (A). The same statements are true for the carpet modulus.

A version of the next lemma can be found in [9, Lemma 2.2] and [6].

Lemma 3.3.2 *Let* $\kappa \geq 1$ *and* I *be a countable index set. Suppose that* $\{B_i\}_{i \in I}$ *is a collection of balls in the plane* $\mathbb{C}$, *and* a_i, $i \in I$, *are non-negative real numbers. Then there exists a constant* $C > 0$ *depending only on* κ *such that*

$$\left\| \sum_{i \in I} a_i \chi_{\kappa B_i} \right\|_2 \leq C \left\| \sum_{i \in I} a_i \chi_{B_i} \right\|_2.$$

Here $\| \cdot \|_2$ denotes the L^2-norm with respect to planar Lebesgue measure.

3.3.3 Existence of Paths

Here we mention some results that provide us with paths that avoid exceptional path families. These paths will be very useful in the proof of injectivity of the uniformizing function f in Sect. 3.8, and also in the proof of the regularity of f^{-1} in Sect. 3.9. A proof of the next proposition can be found in [10, Theorem 3].

Proposition 3.3.3 *Let $\beta \subset \mathbb{C}$ be a path that joins two non-trivial continua E, $F \subset \mathbb{C}$. Consider the distance function $\psi(x) = \text{dist}(x, \beta)$. Then there exists $\delta > 0$ such that for a.e. $s \in (0, \delta)$ there exists a simple path $\beta_s \subset \psi^{-1}(s)$ joining E and F.*

For the next lemma we assume that the carpet S and Ω are as in the assumptions of Theorem 3.1.1, so in particular, the peripheral disks of S are uniformly Ahlfors 2-regular and uniformly quasiround.

Lemma 3.3.4 *Suppose $\beta \subset \Omega$ is a path joining two non-trivial continua E, $F \subset \bar{\Omega}$. Consider the distance function $\psi(x) = \text{dist}(x, \beta)$ and let Γ be a given family in Ω of carpet modulus or conformal modulus equal to zero. Then, there exists $\delta > 0$ such that for a.e. $s \in (0, \delta)$ there exists a simple open path $\beta_s \subset \psi^{-1}(s)$ that lies in Ω, joins the continua E and F, and lies outside the family Γ.*

For a proof see Lemmas 2.4.3 and 2.4.4. We also include some topological facts.

Lemma 3.3.5 (Lemma 2.4.10) *Let $S \subset\subset \mathbb{C}$ be a Sierpiński carpet.*

(a) *For any $x, y \in S$ there exists an open path $\gamma \subset S^\circ$ that joins x and y. Moreover, for each $r > 0$, if y is sufficiently close to x, the path γ can be taken so that $\gamma \subset B(x, r)$.*

(b) *For any two peripheral disks there exists a Jordan curve $\gamma \subset S^\circ$ that separates them. Moreover, γ can be taken to be arbitrarily close to one of them.*

In other words, the conclusion of the first part is that $x, y \in \bar{\gamma}$, but γ does not intersect any peripheral circle ∂Q_i, $i \in \mathbb{N} \cup \{0\}$. As a corollary of the second part of the lemma, we obtain:

Corollary 3.3.6 *Let $S \subset \mathbb{C}$ be a Sierpiński carpet and $\gamma \subset \mathbb{C}$ be a path that connects two distinct peripheral disks of S. Then γ has to intersect S°.*

Lemma 3.3.7 *Let $S \subset \mathbb{C}$ be a Sierpiński carpet and $\gamma \subset \mathbb{C}$ be a non-constant path with $\mathcal{H}^1(\gamma \cap S) = 0$.*

(a) *If $x \in \gamma \cap S^\circ$, then arbitrarily close to x we can find peripheral disks Q_i with $Q_i \cap \gamma \neq \emptyset$.*

(b) *If γ is an open path that does not intersect a peripheral disk Q_{i_0}, $i_0 \in \mathbb{N} \cup \{0\}$, and $x \in \bar{\gamma} \cap \partial Q_{i_0}$, then arbitrarily close to x we can find peripheral disks Q_i, $i \neq i_0$, with $Q_i \cap \gamma \neq \emptyset$.*

The proof is the same as the proof of Lemma 2.4.12, which contains the analogous statement for relative Sierpiński carpets.

We finish the section with a technical lemma regarding the existence of subpaths of a given path that avoid a finite collection of peripheral disks:

Lemma 3.3.8 *Let $J \subset \mathbb{N}$ be a finite index set and $\gamma \subset \Omega$ be an open path with endpoints $x, y \in S$ such that γ does not intersect the closure of any peripheral disk Q_i with $x \in \partial Q_i$ or $y \in \partial Q_i$. Then there exist finitely many open subpaths $\gamma_1, \ldots, \gamma_m$ of γ having endpoints in S with the following properties:*

(i) *γ_i and γ_j intersect disjoint sets of closed peripheral disks for $i \neq j$,*

(ii) *γ_i does not intersect peripheral disks $\overline{Q}_j$, $j \in J$, for all $i \in \{1, \ldots, m\}$,*

(iii) *γ_1 starts at $x_1 = x$, γ_m terminates at $y_m = y$, and in general the path γ_i starts at x_i and terminates at y_i such that for each $i \in \{1, \ldots, m-1\}$ we either have*

- *$y_i = x_{i+1}$, i.e., γ_i and γ_{i+1} have a common endpoint, or*
- *$y_i, x_{i+1} \in \partial Q_{j_i}$ for some $j_i \in \mathbb{N}$, i.e., γ_i and γ_{i+1} have an endpoint on some peripheral circle ∂Q_{j_i}.*

The peripheral disks $\overline{Q}_{j_i}$ that arise from the second case are distinct and they are all intersected by the original curve γ.

This is a variant of Lemma 2.5.10 and its proof is omitted.

3.4   The Function u

From this section until Sect. 3.9 the standing assumptions are that we are given a carpet $S \subset \overline{\Omega}$ of area zero with peripheral disks $\{Q_i\}_{i \in \mathbb{N}}$ that are uniformly Ahlfors 2-regular and uniformly quasiround, and with outer peripheral circle $\partial Q_0 = \partial \Omega$. This is precisely the setup in Chap. 2, where the theory of carpet-harmonic functions is developed. We will use this theory in order to define the real part u of the uniformizing function f.

3.4.1   Background on Carpet-Harmonic Functions

Here we include some definitions and background on carpet-harmonic functions. More details can be found in Chap. 2.

Definition 3.4.1 Let $g \colon S \cap \Omega \to \widehat{\mathbb{R}}$ be an extended function. We say that the sequence of non-negative weights $\{\lambda(Q_i)\}_{i \in \mathbb{N}}$ is an *upper gradient* for g if there exists an exceptional family Γ_0 of paths in Ω with $\text{mod}(\Gamma_0) = 0$ such that for all paths $\gamma \subset \Omega$ with $\gamma \notin \Gamma_0$ and $x, y \in \gamma \cap S$ we have $g(x), g(y) \neq \pm\infty$ and

$$|g(x) - g(y)| \leq \sum_{i: Q_i \cap \gamma \neq \emptyset} \lambda(Q_i).$$

Note that this definition differs from the classical definition of upper gradients in metric spaces (given in Chap. 1), treated in [38] and [22]. In the classical definition, one uses line integrals (instead of sums) over paths that are entirely contained in the metric space. However, here, we use paths that are not contained in the carpet and intersect the peripheral disks Q_i in the complement of the carpet. In fact, the presence of ambient space is important, since "most" of the paths do not lie in the carpet S, but meet infinitely many peripheral disks Q_i. We also remark that our notation here differs slightly from Definition 2.5.11, where a *relative carpet* S does not contain $\partial\Omega$, and functions are defined on S. Since here S contains $\partial\Omega$, we write $S \cap \Omega$ here as the domain of g.

For a function $g : S \cap \Omega \to \widehat{\mathbb{R}}$ and a peripheral disk Q_i, $i \in \mathbb{N}$, we define

$$M_{Q_i}(g) = \sup_{x \in \partial Q_i} g(x) \quad \text{and} \quad m_{Q_i}(g) = \inf_{x \in \partial Q_i} g(x).$$

If at least one of the above quantities is finite, then we define

$$\operatorname*{osc}_{Q_i}(g) = M_{Q_i}(g) - m_{Q_i}(g).$$

Note here that we do *not* define the above quantities for the outer peripheral circle $\partial Q_0 = \partial\Omega$, which is regarded as the "boundary" of the carpet.

Definition 3.4.2 Let $g : S \cap \Omega \to \widehat{\mathbb{R}}$ be an extended function. We say that g lies in the *Sobolev space* $\mathcal{W}_{\mathrm{loc}}^{1,2}(S)$ if for every ball $B \subset\subset \Omega$ we have

$$\sum_{i \in I_B} M_{Q_i}(g)^2 \operatorname{diam}(Q_i)^2 < \infty, \tag{3.6}$$

$$\sum_{i \in I_B} \operatorname*{osc}_{Q_i}(g)^2 < \infty, \tag{3.7}$$

and $\{\operatorname{osc}_{Q_i}(g)\}_{i \in \mathbb{N}}$ is an upper gradient for g. If the above conditions hold for the full sums over $i \in \mathbb{N}$ then we say that g lies in the Sobolev space $\mathcal{W}^{1,2}(S)$.

Recall here that $I_B = \{i \in \mathbb{N} : Q_i \cap B \neq \emptyset\}$. Part of the definition is that $\operatorname{osc}_{Q_i}(g)$ is defined for all $i \in \mathbb{N}$, and in particular $M_{Q_i}(g), m_{Q_i}(g)$ are finite. The space $\mathcal{W}^{1,2}(S)$ contains Lipschitz functions on S, and also coordinate functions of restrictions on S of quasiconformal maps $g : \mathbb{C} \to \mathbb{C}$; see Sect. 2.5.3.

For a set $V \subset \Omega$ and $g \in \mathcal{W}_{\mathrm{loc}}^{1,2}(S)$ define the *Dirichlet energy functional*

$$D_V(g) = \sum_{i \in I_V} \operatorname*{osc}_{Q_i}(g)^2 \in [0, \infty].$$

We remark here that the outer peripheral disk Q_0 is never used in Dirichlet energy calculations, and the summations are always over subsets of $\{Q_i\}_{i\in\mathbb{N}}$. If $V = \Omega$ we will often omit the subscript and write $D(g)$ instead of $D_\Omega(g)$.

Definition 3.4.3 A function $u \in W^{1,2}_{loc}(S)$ is *carpet-harmonic* if for every open set $V \subset\subset \Omega$ and every $\zeta \in W^{1,2}(S)$ that is supported on V we have

$$D_V(u) \le D_V(u + \zeta).$$

For each $g \in W^{1,2}(S)$ there exists a family of *good paths* G in Ω that contains almost every path (i.e., the paths of Ω that do not lie in G have carpet modulus equal to zero) with the following properties

1. $\mathcal{H}^1(\gamma \cap S) = 0$,
2. $\sum_{i:Q_i\cap\gamma\neq\emptyset} \mathrm{osc}_{Q_i}(g) < \infty$, and
3. the upper gradient inequality as in Definition 3.4.1 holds along every subpath of γ.

A point $x \in S$ is *accessible by a path* $\gamma_0 \in G$ if there exists an open subpath γ of γ_0 with $x \in \overline{\gamma}$ and γ does not meet the peripheral disk Q_{i_0} whenever $x \in \partial Q_{i_0}$, $i_0 \in \mathbb{N}$; see Fig. 2.1. Note that x can lie on $\partial\Omega$. See Sect. 2.5.1 for a more detailed discussion on good paths and accessible points.

Finally we require a lemma that allows the "gluing" of Sobolev functions and will be useful for variational arguments; see Proposition 2.5.27 and Lemma 2.5.28 for the proof.

Lemma 3.4.4 *If* $\phi, \psi \in W^{1,2}(S)$ *and* $a, b \in \mathbb{R}$, *then the following functions also lie in the Sobolev space* $W^{1,2}(S)$:

(a) $a\phi + b\psi$, *with* $\mathrm{osc}_{Q_i}(a\phi + b\psi) \le |a|\,\mathrm{osc}_{Q_i}(\phi) + |b|\,\mathrm{osc}_{Q_i}(\psi)$,
(b) $|\phi|$, *with* $\mathrm{osc}_{Q_i}(|\phi|) \le \mathrm{osc}_{Q_i}(\phi)$,
(c) $\phi \vee \psi := \max(\phi, \psi)$, *with* $\mathrm{osc}_{Q_i}(\phi \vee \psi) \le \max\{\mathrm{osc}_{Q_i}(\phi), \mathrm{osc}_{Q_i}(\psi)\}$,
(d) $\phi \wedge \psi := \min(\phi, \psi)$, *with* $\mathrm{osc}_{Q_i}(\phi \wedge \psi) \le \max\{\mathrm{osc}_{Q_i}(\phi), \mathrm{osc}_{Q_i}(\psi)\}$, *and*
(e) $\phi \cdot \psi$, *provided that* ϕ *and* ψ *are bounded.*

Furthermore, if $V \subset \Omega$ *is an open set with* $S \cap \Omega \cap \partial V \neq \emptyset$, *and* $\phi = \psi$ *on* $S \cap \Omega \cap \partial V$, *then* $\phi\chi_{S\cap V} + \psi\chi_{S\setminus V} \in W^{1,2}(S)$.

3.4.2   The Free Boundary Problem

We mark four points on $\partial\Omega$ that determine a quadrilateral, i.e., a homeomorphic image of a rectangle, with closed sides $\Theta_1, \ldots, \Theta_4$, enumerated in a counter-clockwise fashion. Here Θ_1 and Θ_3 are opposite sides.

Consider a function $g \in W^{1,2}(S)$. Recall from Definition 3.4.2 that g is only defined in $S \cap \Omega$ and not in $\partial\Omega$. However, one can always define boundary values of g on $\partial\Omega$; see Sect. 2.6.2 for more details. In this chapter, all functions $g \in W^{1,2}(S)$

that we are going to use will actually be continuous up to $\partial\Omega$, so their boundary values are unambiguously defined and we do not need to resort to the theory of Chap. 2. If $g(x) = 0$ for all points $x \in \Theta_1$ and also $g(x) = 1$ for all points $x \in \Theta_3$, we say that g is *admissible (for the free boundary problem)*.

Theorem 3.4.5 (Theorem 2.7.11) *There exists a unique carpet-harmonic function* $u \colon S \to \mathbb{R}$ *that minimizes the Dirichlet energy* $D_\Omega(g)$ *over all admissible functions* $g \in \mathcal{W}^{1,2}(S)$. *The function* u *is continuous up to the boundary* $\partial\Omega$ *and has boundary values* $u = 0$ *on* Θ_1 *and* $u = 1$ *on* Θ_3.

For an open set $V \subset \mathbb{C} \setminus (\Theta_1 \cup \Theta_3)$ define

$$\partial_* V = \partial V \cap S.$$

The open arcs $\Theta_2, \Theta_4 \subset \partial\Omega$ are not considered as boundary arcs for the free boundary problem since there is no boundary data present on them. With this in mind, we now state the maximum principle for the minimizer u:

Theorem 3.4.6 (Theorem 2.7.12) *Let V be an open set with $V \subset \mathbb{C} \setminus (\Theta_1 \cup \Theta_3)$. Then*

$$\sup_{x \in S \cap \overline{V}} u(x) = \sup_{x \in \partial_* V} u(x) \quad \text{and} \quad \inf_{x \in S \cap \overline{V}} u(x) = \inf_{x \in \partial_* V} u(x).$$

The standard maximum principle would state that the extremal values of u on an open set V are attained on ∂V. Our stronger statement states that the extremal values could be attained at the part of ∂V that is disjoint from the interiors of the "free" arcs Θ_2 and Θ_4. However, extremal values could still be attained at Θ_1 or Θ_3, and this is the reason that we look at sets $V \subset \mathbb{C} \setminus (\Theta_1 \cup \Theta_3)$. See Fig. 2.5.

Next, we consider a variant of Lemma 2.9.5, whose proof follows immediately from an application of the upper gradient inequality, together with the maximum principle.

Lemma 3.4.7 *Consider a ball $B(x, r) \subset \Omega$, with $B(x, cr) \subset \Omega$ for some $c > 1$. Then for a.e. $s \in [1, c]$ we have*

$$\operatorname{diam}(u(B(x, r) \cap S)) \leq \operatorname{diam}(u(B(x, sr) \cap S)) \leq \sum_{i : Q_i \cap \partial B(x, sr) \neq \emptyset} \operatorname*{osc}_{Q_i}(u).$$

The function u will be the real part of the uniformizing function f. It will be very convenient to have a continuous extension of u to $\overline{\Omega}$ that satisfies the maximum principle:

Proposition 3.4.8 *There exists a continuous extension* $\tilde{u}\colon \overline{\Omega} \to \mathbb{R}$ *of* u *such that for every open set* $V \subset \mathbb{C} \setminus (\Theta_1 \cup \Theta_3)$ *we have*

$$\sup_{x \in \overline{V} \cap \overline{\Omega}} \tilde{u}(x) = \sup_{x \in \partial V \cap \overline{\Omega}} \tilde{u}(x) \quad and \quad \inf_{x \in \overline{V} \cap \overline{\Omega}} \tilde{u}(x) = \inf_{x \in \partial V \cap \overline{\Omega}} \tilde{u}(x).$$

In fact, $\tilde{u}$ *can be taken to be harmonic in the classical sense inside each peripheral disk* Q_i, $i \in \mathbb{N}$.

Proof For each peripheral disk Q_i, $i \in \mathbb{N}$, we consider the Poisson extension $\tilde{u}\colon \overline{Q}_i \to \mathbb{R}$ of u. This is obtained by mapping conformally the Jordan region Q_i to the unit disk $\mathbb{D}$ and taking the Poisson extension there. The function $\tilde{u}$ is harmonic in Q_i and continuous up to the boundary ∂Q_i. Furthermore,

$$\operatorname{diam}(\tilde{u}(Q_i)) = \operatorname{diam}(\tilde{u}(\partial Q_i)) = \operatorname*{osc}_{Q_i}(u), \tag{3.8}$$

where the latter is defined after Definition 3.4.1.

To show that the extension $\tilde{u}\colon \overline{\Omega} \to \mathbb{R}$ is continuous, we argue by contradiction and suppose that there exists a sequence $\{x_n\}_{n \in \mathbb{N}} \subset \overline{\Omega}$ with $x_n \to x \in \overline{\Omega}$, but $|\tilde{u}(x_n) - \tilde{u}(x)| \geq \varepsilon$ for some $\varepsilon > 0$ and all $n \in \mathbb{N}$. If $x_n \in S$ for infinitely many n, then we obtain a contradiction, by the continuity of u in S. If x_n lies in some peripheral disk $\overline{Q}_{i_0}$ for infinitely many n then we also get a contradiction, by the continuity of $\tilde{u}$ on $\overline{Q}_{i_0}$. We, thus, assume that $x_n \in \overline{Q}_{i_n}$ where Q_{i_n} are distinct peripheral disks, so we necessarily have $x \in S$. Let $y_n \in \partial Q_{i_n}$. By the local connectedness of S we have $\operatorname{diam}(Q_{i_n}) \to 0$, thus $y_n \to x$ and $u(y_n) \to u(x)$. Since $\sum_{i \in \mathbb{N}} \operatorname{osc}_{Q_i}(u)^2 < \infty$ it follows that $\operatorname{osc}_{Q_{i_n}}(u) \to 0$. Combining these with (3.8) we obtain

$$\varepsilon \leq |\tilde{u}(x_n) - \tilde{u}(x)| \leq |\tilde{u}(x_n) - \tilde{u}(y_n)| + |u(y_n) - u(x)|$$

$$\leq \operatorname*{osc}_{Q_{i_n}}(u) + |u(y_n) - u(x)|$$

Letting $n \to \infty$ yields, again, a contradiction.

Finally, we check the maximum principle. Trivially, we have

$$\sup_{x \in \overline{V} \cap \overline{\Omega}} \tilde{u}(x) \geq \sup_{x \in \partial V \cap \overline{\Omega}} \tilde{u}(x) =: M,$$

so it suffices to show the reverse inequality. If there exists $z \in V \cap \overline{\Omega}$ with $\tilde{u}(z) > M$, then we claim that there actually exists $w \in S \cap V$ with $u(w) = \tilde{u}(w) > M$. We assume this for the moment. Since $\partial V \cap \overline{\Omega} \supset \partial_* V$, we have

$$\sup_{x \in S \cap \overline{V}} u(x) \geq u(w) > M = \sup_{x \in \partial V \cap \overline{\Omega}} \tilde{u}(x) \geq \sup_{x \in \partial_* V} \tilde{u}(x) = \sup_{x \in \partial_* V} u(x),$$

which contradicts the maximum principle in Theorem 3.4.6. The statement for the infimum is proved similarly.

We now prove our claim. If $z \in S$ then we set $w = z$ and there is nothing to show, so we assume that $z \in Q_{i_0}$ for some $i_0 \in \mathbb{N}$. The maximum principle of the harmonic function $\tilde{u}\big|_{Q_{i_0}}$ implies that there exists

$$w \in \partial(Q_{i_0} \cap V) \subset (\partial Q_{i_0} \cap V) \cup \partial V \subset (S \cap V) \cup \partial V$$

with $\tilde{u}(w) > M$; here it is crucial that Q_{i_0} and V are open sets. However, we cannot have $w \in \partial V$, since $\tilde{u}(w) > M = \sup_{x \in \partial V \cap \overline{\Omega}} \tilde{u}(x)$. It follows that $w \in S \cap V$, as desired. $\qquad\square$

3.5 The Level Sets of u

We study the level sets of u and of its extension $\tilde{u}$. One of our goals is to show that for a.e. $t \in [0, 1]$ the level set $\tilde{u}^{-1}(t)$ is a simple curve that joins Θ_2 to Θ_4. This strategy is also followed in [33, Sect. 6], although the technical details are substantially different. Using these curves we will define the "conjugate" function v of u in the next section.

For $0 \le s < t \le 1$ define

$$A_{s,t} = \tilde{u}^{-1}((s, t))$$

and for $0 \le t \le 1$ define $\alpha_t = \tilde{u}^{-1}(t)$. For these level sets we have the following.

Proposition 3.5.1 *For all $0 \le s < t \le 1$ the sets α_t and $A_{s,t}$ are connected, simply connected and they join the sides Θ_2 and Θ_4 of the quadrilateral Ω. Furthermore, the intersections of α_t, $A_{s,t}$ with Θ_2, Θ_4 are all connected. Finally, α_t does not separate the plane and $\overline{A}_{s,t}$ does not separate the plane if α_t and α_s have empty interior.*

This is proved in the same way as [33, Lemma 6.3], but we include a sketch of it here for the sake of completeness.

Proof We prove the statement for the set $A := A_{s,t}$, which is relatively open in $\overline{\Omega}$. The claims for α_t are proved very similarly, observing also that $\alpha_t = \bigcap_{h>0} A_{t-h,t+h}$ for $0 < t < 1$.

We first show that each component V of A is simply connected. If there exists a simple loop $\gamma \subset V$ that is not null-homotopic in V, then γ bounds a region W that is not contained in A. However, $\partial W \cap \overline{\Omega} = \gamma$ is contained in A, so we have $s < \tilde{u}(x) < t$ for all $x \in \partial W \cap \overline{\Omega}$. The maximum principle in Theorem 3.4.8 implies that this also holds in W, a contradiction.

Let V be a component of A. Then V has to intersect at least one of the sides Θ_2 and Θ_4. Indeed, if this was not the case, then on the connected set $\partial V \subset \Omega$ we would either have $\tilde{u} \equiv s$ or $\tilde{u} \equiv t$. The maximum principle in Theorem 3.4.8 implies that $\tilde{u}$ is a constant on V, equal to either s or t, but this clearly contradicts the fact that $V \subset A = \tilde{u}^{-1}((s, t))$. Without loss of generality we assume that $V \cap \Theta_2 \neq \emptyset$.

The intersection $V \cap \Theta_2$ must be a connected set. Indeed, if this failed, then we would be able to find a simple arc $\gamma \subset V$ that connects two distinct components of $V \cap \Theta_2$. Since $\tilde{u} \in (s, t)$ on γ, it follows again by the maximum principle that the same is true in the region bounded by γ and Θ_2. This is a contradiction. Note that here the maximum principle is applied to an open set $W \subset \mathbb{C} \setminus (\Theta_1 \cup \Theta_3)$ bounded by the concatenation of γ with an arc $\beta \subset \mathbb{C} \setminus \overline{\Omega}$ that connects the endpoints of γ.

Our next claim is that V intersects Θ_4. We argue by contradiction, so suppose that the boundary of V rel. $\overline{\Omega}$ consists of a single component Y. On Y we must have $\tilde{u} \equiv s$ or $\tilde{u} \equiv t$ and only one of them is possible by the connectedness of Y. In either case, $\tilde{u}$ would have to be constant in V by the maximum principle, and this is a contradiction as in the previous paragraph.

Suppose now there exist two distinct components $V_1, V_3 \subset A$. Since both of them separate the sides Θ_1 and Θ_3, there exists some $x \in \overline{\Omega} \setminus A$ "between" V_1 and V_3, i.e., A separates x from both Θ_1 and Θ_3. The maximum principle applied to the region containing x, bounded by parts of the boundaries of V_1 and V_3, is again contradicted.

For our final claim, suppose that $\mathbb{C} \setminus \overline{A}$ has a bounded component V. Note that V is simply connected by the connectedness of $\overline{A}$, and hence the boundary ∂V of V is connected. The set ∂V cannot contain an arc of $\partial \Omega$. Indeed, otherwise we would be able to connect V with the unbounded component of $\mathbb{C} \setminus \overline{A}$ outside $\overline{A}$, a contradiction. Hence, $\partial V \cap \partial \Omega$ is a totally disconnected set, and each point of $\partial V \cap \partial \Omega$ can be approximated by points in $\partial V \cap \Omega$. On each component of $\partial V \setminus \partial \Omega \subset \partial A \cap \Omega$ we necessarily have $\tilde{u} \equiv s$ or $\tilde{u} \equiv t$. By continuity it follows that on each point of $\partial V \cap \partial \Omega$ the function $\tilde{u}$ has the value s or t. Since ∂V is connected, we have $\tilde{u} \equiv s$ or $\tilde{u} \equiv t$ on ∂V. The maximum principle implies that $V \subset \alpha_s$ or $V \subset \alpha_t$, but this contradicts the assumption that the level sets α_s and α_t have empty interior. $\square$

Next, we prove the following theorem.

Theorem 3.5.2 *For a.e.* $t \in (0, 1)$ *we have*

$$\sum_{i: Q_i \cap \alpha_t \neq \emptyset} \operatorname*{osc}_{Q_i}(u) = D(u) = \sum_{i \in \mathbb{N}} \operatorname*{osc}_{Q_i}(u)^2.$$

The proof will follow from Propositions 3.5.3 and 3.5.5 below.

Proposition 3.5.3 *For a.e.* $t \in (0, 1)$ *we have*

$$\sum_{i: Q_i \cap \alpha_t \neq \emptyset} \operatorname*{osc}_{Q_i}(u) \geq D(u).$$

Proof Let $t \in (0, 1)$ be a value that is not the maximum or minimum value of u on any ∂Q_i, $i \in \mathbb{N}$. There are countably many such values that we exclude. Then

$Q_i \cap \alpha_{t \pm h} \neq \emptyset$ for all small $h > 0$, whenever $Q_i \cap \alpha_t \neq \emptyset$. Indeed, if $x \in Q_i \cap \alpha_t$, and $\tilde{u}$ is non-constant in Q_i, then by harmonicity $\tilde{u}$ must attain, near x, values larger than t and smaller than t. We fix a sufficiently small $h > 0$ and define F_h to be the family of indices $i \in \mathbb{N}$ such that $Q_i \cap \alpha_{t+h} \neq \emptyset$ and $Q_i \cap \alpha_{t-h} \neq \emptyset$. Note that F_h is contained in $\{i \in \mathbb{N} : Q_i \cap \alpha_t \neq \emptyset\}$ by the connectedness of Q_i and the continuity of u. In fact, by our previous remark, F_h increases to $\{i \in \mathbb{N} : Q_i \cap \alpha_t \neq \emptyset\}$ as $h \to 0$. Hence, we have

$$\sum_{i \in F_h} \operatorname*{osc}_{Q_i}(u) \to \sum_{i : Q_i \cap \alpha_t \neq \emptyset} \operatorname*{osc}_{Q_i}(u) \tag{3.9}$$

as $h \to 0$. Also, define $N_h = \{i \in \mathbb{N} : Q_i \cap A_{t-h,t+h} \neq \emptyset\} \setminus F_h$.

Now, consider the function

$$g(x) = \begin{cases} 0, & u(x) \leq t - h \\ \frac{u(x) - (t-h)}{2h}, & t - h < u(x) < t + h \\ 1, & u(x) \geq t + h. \end{cases}$$

The function g lies in the Sobolev space $\mathcal{W}^{1,2}(S)$ as follows from Lemma 3.4.4, and furthermore we have

$$\operatorname*{osc}_{Q_i}(g) \begin{cases} = 1, & i \in F_h \\ \leq \operatorname*{osc}_{Q_i}(u)/2h, & i \in N_h \\ = 0, & i \notin F_h \cup N_h. \end{cases}$$

Since $g = 0$ on Θ_1 and $g = 1$ on Θ_3, the function g is admissible for the free boundary problem.

Hence, for all $s \in [0, 1]$ the function $(1 - s)u + sg$ is also admissible for the free boundary problem, so $D(u) \leq D((1 - s)u + sg)$ by the harmonicity of u. Lemma 3.4.4(a) implies that

$$D(u) \leq \sum_{i \in \mathbb{N}} ((1 - s) \operatorname*{osc}_{Q_i}(u) + s \operatorname*{osc}_{Q_i}(g))^2$$

$$= (1 - s)^2 D(u) + 2(1 - s)s \sum_{i \in \mathbb{N}} \operatorname*{osc}_{Q_i}(u) \operatorname*{osc}_{Q_i}(g) + s^2 D(g).$$

This simplifies to

$$D(u) \leq (1 - s) \sum_{i \in \mathbb{N}} \operatorname*{osc}_{Q_i}(u) \operatorname*{osc}_{Q_i}(g) + \frac{s}{2}(D(u) + D(g)).$$

Letting $s \to 0$, we obtain

$$D(u) \leq \sum_{i \in \mathbb{N}} \operatorname*{osc}_{Q_i}(u) \operatorname*{osc}_{Q_i}(g), \tag{3.10}$$

thus,

$$D(u) \leq \sum_{i \in F_h} \operatorname*{osc}_{Q_i}(u) + \frac{1}{2h} \sum_{i \in N_h} \operatorname*{osc}_{Q_i}(u)^2. \tag{3.11}$$

By (3.9), it suffices to prove that $\frac{1}{2h} \sum_{i \in N_h} \operatorname{osc}_{Q_i}(u)^2 \to 0$ as $h \to 0$, for a.e. $t \in [0, 1]$. This will follow from the next lemma.

Lemma 3.5.4 *Let* $\{h(Q_i)\}_{i \in \mathbb{N}}$ *be a sequence of non-negative numbers that is summable, i.e.,* $\sum_{i \in \mathbb{N}} h(Q_i) < \infty$. *For each* $i \in \mathbb{N}$ *consider points* $p_i, q_i \in \partial Q_i$ *such that* $u(p_i) = m_{Q_i}(u) = \min_{\partial Q_i} u$ *and* $u(q_i) = M_{Q_i}(u) = \max_{\partial Q_i} u$. *Define a measure* μ *on* $\mathbb{C}$ *by*

$$\mu = \sum_{i \in \mathbb{N}} h(Q_i)(\delta_{p_i} + \delta_{q_i}),$$

where δ_x *is a Dirac mass at* x. *Then for the pushforward measure* $\lambda := u_* \mu$ *on* $\mathbb{R}$ *we have*

$$\lim_{h \to 0} \frac{\lambda((t - h, t + h))}{2h} = \lim_{h \to 0} \frac{1}{2h} \sum_{i \in \mathbb{N}} h(Q_i)(\chi_{p_i \in A_{t-h,t+h}} + \chi_{q_i \in A_{t-h,t+h}}) = 0$$

for a.e. $t \in [0, 1]$.

The proof of the lemma is an immediate consequence of the fact that the measure λ has no absolutely continuous part; see [18, Theorem 3.22, p. 99]. We now explain how to use the lemma in order to derive that the term $\frac{1}{2h} \sum_{i \in N_h} \operatorname{osc}_{Q_i}(u)^2$ in (3.11) converges to 0 as $h \to 0$, for a.e. $t \in [0, 1]$.

We first refine our choice of $t \in (0, 1)$ such that the conclusion of Lemma 3.5.4 is true for $h(Q_i) := \operatorname{osc}_{Q_i}(u)^2$. Recall that initially we only excluded countably many values of t. If $i \in N_h$ then we have $u(p_i) \in (t-2h, t+2h)$ or $u(q_i) \in (t-2h, t+2h)$. To see this first note that $\partial Q_i \cap A_{t-h,t+h} \neq \emptyset$ in this case by the connectedness of $A_{t-h,t+h}$ from Proposition 3.5.1. Thus $M_{Q_i}(u) = u(q_i) \geq t - h$. Since $i \notin F_h$, without loss of generality assume that $Q_i \cap \alpha_{t+h} = \emptyset$. By continuity, the maximum of u on ∂Q_i cannot exceed $t + h$, so $u(q_i) \in [t - h, t + h] \subset (t - 2h, t + 2h)$. Therefore, $\sum_{i \in N_h} \operatorname{osc}_{Q_i}(u)^2 \leq \lambda((t-2h, t+2h))$ and this completes the proof. □

Proposition 3.5.5 *For a.e.* $t \in (0, 1)$ *we have*

$$\sum_{i : Q_i \cap \alpha_t \neq \emptyset} \operatorname*{osc}_{Q_i}(u) \leq D(u).$$

Proof Again we choose a $t \in (0, 1)$ that is not a maximum or minimum value of u on any ∂Q_i, $i \in \mathbb{N}$.

We fix $\delta, h > 0$ and we define a Sobolev function $g \in \mathcal{W}^{1,2}(S)$ with $g = 0$ on Θ_1 and $g = 1 + \delta(1 - 2h)$ on Θ_3 as follows:

$$
g(x) = \begin{cases} (1+\delta)u(x), & u(x) \le t - h \\ u(x) + c_1, & t - h < u(x) < t + h \\ (1+\delta)u(x) + c_2, & u(x) \ge t + h \end{cases}
$$

where the constants c_1, c_2 are chosen so that g is continuous. It easy to see that $c_1 = \delta(t - h)$ and $c_2 = -2\delta h$. Note that g can be written as

$$
g = [((1+\delta)u) \wedge (u + c_1)] \vee ((1+\delta)u + c_2), \tag{3.12}
$$

which shows that $g \in \mathcal{W}^{1,2}(S)$, according to Lemma 3.4.4. Consider the index sets $F_h = \{i \in \mathbb{N} : Q_i \cap \alpha_{t+h} \ne \emptyset \text{ and } Q_i \cap \alpha_{t-h} \ne \emptyset\}$ and $N_h = \{i \in \mathbb{N} : Q_i \not\subset A_{t-h,t+h}\} \setminus F_h$. Observe that

$$
\underset{Q_i}{\operatorname{osc}}(g) \begin{cases} \le (1+\delta)\operatorname{osc}_{Q_i}(u) - 2\delta h, & i \in F_h \\ \le (1+\delta)\operatorname{osc}_{Q_i}(u), & i \in N_h \\ = \operatorname{osc}_{Q_i}(u), & i \notin F_h \cup N_h. \end{cases}
$$

Indeed, for the first inequality note that for $i \in F_h$, the maximum of g on ∂Q_i has to be attained at a point x with $u(x) \ge t + h$, and the minimum is attained at a point y with $u(y) \le t - h$. Hence, $g(x) - g(y) = (1+\delta)(u(x) - u(y)) - 2\delta h \le (1+\delta)\operatorname{osc}_{Q_i}(u) - 2\delta h$. The second inequality holds for all $i \in \mathbb{N}$ and is a crude estimate, based on (3.12) and Lemma 3.4.4; here it is crucial that $\delta > 0$ so that $\operatorname{osc}_{Q_i}(u) \le (1+\delta)\operatorname{osc}_{Q_i}(u)$. The third equality is immediate, since $t - h \le u \le t + h$ and thus $g = u + c_1$ on ∂Q_i, whenever $i \notin F_h \cup N_h$.

The function $g/(1 + \delta(1 - 2h))$ is admissible for the free boundary problem, and testing the minimizing property of u against $g/(1 + \delta(1 - 2h))$ as in the proof of Proposition 3.5.3 (see (3.10)) we obtain

$$
D(u) \le \frac{1}{1 + \delta(1 - 2h)} \sum_{i \in \mathbb{N}} \underset{Q_i}{\operatorname{osc}}(u) \underset{Q_i}{\operatorname{osc}}(g).
$$

This implies that

$$
D(u) \le \frac{1}{1 + \delta(1 - 2h)}\bigg((1+\delta)\sum_{i \in F_h} \underset{Q_i}{\operatorname{osc}}(u)^2 - 2\delta h \sum_{i \in F_h} \underset{Q_i}{\operatorname{osc}}(u)
$$

$$
+ (1+\delta)\sum_{i \in N_h} \underset{Q_i}{\operatorname{osc}}(u)^2 + \sum_{i \notin F_h \cup N_h} \underset{Q_i}{\operatorname{osc}}(u)^2 \bigg).
$$

Manipulating the expression yields

$$\sum_{i \in F_h} \underset{Q_i}{\mathrm{osc}}(u) + \frac{1}{2h} \sum_{i \notin F_h \cup N_h} \underset{Q_i}{\mathrm{osc}}(u)^2 \le D(u). \tag{3.13}$$

By the choice of t we have

$$\sum_{i \in F_h} \underset{Q_i}{\mathrm{osc}}(u) \to \sum_{i : Q_i \cap \alpha_t \neq \emptyset} \underset{Q_i}{\mathrm{osc}}(u)$$

as $h \to 0$. Moreover, if $i \notin F_h \cup N_h$, then $\overline{Q}_i \subset A_{t-2h,t+2h}$, so

$$\sum_{i \notin F_h \cup N_h} \underset{Q_i}{\mathrm{osc}}(u)^2 \le \lambda((t - 2h, t + 2h)),$$

where λ is as in Lemma 3.5.4 and $h(Q_i) := \mathrm{osc}_{Q_i}(u)^2$. By the lemma, it follows that

$$\frac{1}{2h} \sum_{i \notin F_h \cup N_h} \underset{Q_i}{\mathrm{osc}}(u)^2 \to 0$$

as $h \to 0$ for a.e. $t \in [0, 1]$. This, together with (3.13) yields the conclusion.   □

Another important topological property of the level sets of u is the following:

Lemma 3.5.6 *For a.e. $t \in [0, 1]$ and for all $i \in \mathbb{N}$ the intersection $u^{-1}(t) \cap \partial Q_i$ contains at most two points. Furthermore, for a.e. $t \in [0, 1]$ the intersection $u^{-1}(t) \cap \partial \Omega$ contains exactly two points, one in Θ_2 and one in Θ_4.*

The proof is based on the following elementary lemma, which is the one-dimensional version of Sard's theorem:

Lemma 3.5.7 *Let $f : \mathbb{R} \to \mathbb{R}$ be an arbitrary function. Then the set of local maximum and local minimum values of f is at most countable.*

We will use the lemma now and provide a proof right afterwards.

Proof of Lemma 3.5.6 Note that $\partial \Omega$, ∂Q_i can be identified with $\mathbb{R}/\mathbb{Z}$. We consider a level $t \in (0, 1)$ that is not a local maximum or local minimum value of u on $\partial \Omega$ and on any peripheral circle ∂Q_i, $i \in \mathbb{N}$. This implies that for each point $x \in u^{-1}(t) \cap \partial Q_i$ there exist, arbitrarily close to x, points $x_+, x_- \in \partial Q_i$ with $u(x_+) > t$ and $u(x_-) < t$.

By Proposition 3.5.1, $u^{-1}(t)$ intersects Θ_2 at a connected set. Since t is not a local maximum or local minimum value of u on $\partial \Omega$, it follows that $u^{-1}(t) \cap \Theta_2$ cannot be an arc, so it has to be a point. Similarly, $u^{-1}(t) \cap \Theta_4$ is a singleton. Furthermore, with the same reasoning, for each $i \in \mathbb{N}$ the intersection $u^{-1}(t) \cap \partial Q_i$ is a totally disconnected set. Let ∂Q_i, $i \in \mathbb{N}$, be an arbitrary peripheral circle, intersected by $u^{-1}(t)$. We now split the proof in two parts.

Step 1 There exist continua $C_2, C_4 \subset \tilde{u}^{-1}(t)$ that connect Θ_2, Θ_4 to ∂Q_i, respectively, with $C_2 \cap \partial Q_i = \{x_2\}$ and $C_4 \cap \partial Q_i = \{x_4\}$ for some points $x_2, x_4 \in \partial Q_i$. We provide details on how to obtain these continua.

Let $C_2 \subset \tilde{u}^{-1}(t)$ be a minimal continuum that connects ∂Q_i to Θ_2, and $C_4 \subset \tilde{u}^{-1}(t)$ be a minimal continuum that connects ∂Q_i to Θ_4. Here, a continuum C joining two sets E and F is minimal if any compact proper subset of C is either disconnected or it does not connect E and F. The existence of the continua C_2 and C_4 follows from Zorn's lemma, because the intersection of a chain of continua connecting two compact sets is again a continuum connecting these compact sets; see [46, Theorem 28.2] and the proof of [46, Theorem 28.4]. We next show that C_2 intersects ∂Q_i at a single point x_2 and C_4 intersects ∂Q_i at a single point x_4.

First note that $C_2 \cap Q_i = \emptyset$. Otherwise, $C_2 \setminus Q_i$ has to be disconnected by the minimality of C_2, so there exists a compact component W of $C_2 \setminus Q_i$ that intersects Θ_2. By the minimality of C_2, W cannot intersect ∂Q_i, so it has a positive distance from it. The component W is the intersection of all rel. clopen subsets of $C_2 \setminus Q_i$ that contain it; see [11, Corollary 1.34] or [34, p. 304]. Let $U \supset W$ be such a clopen set, very close to W, so that $U \cap \partial Q_i = \emptyset$. Then $C_2 = U \cup (C_2 \setminus U)$, where U and $C_2 \setminus U$ are non-empty and rel. closed in C_2. This contradicts the connectedness of C_2, and completes the proof that $C_2 \cap Q_i = \emptyset$.

Now, assume that $C_2 \cap \partial Q_i$ contains two points, x and y. We connect these points by a simple path $\gamma \subset Q_i$. Then, we claim that $\overline{\Omega} \setminus (C_2 \cup \gamma)$ has a component $V \subset \Omega$ such that $\partial V \subset C_2 \cup \gamma$, and such that $\overline{V}$ contains an arc $\beta \subset \partial Q_i$ between x and y. Assume the claim for the moment. On $\partial_* V = \partial V \cap S$ we have $u \equiv t$, so by the maximum principle in Theorem 3.4.6 we obtain that $u \equiv t$ on $\overline{V} \cap S$. However, this implies that $u \equiv t$ on β and this contradicts the fact that $u^{-1}(t) \cap \partial Q_i$ is totally disconnected. Thus, indeed $C_2 \cap \partial Q_i \subset u^{-1}(t)$ contains precisely one point, x_2. The same is true for $C_4 \cap \partial Q_i = \{x_4\}$.

Now we prove the claim. Note that γ separates Q_i in two open "pieces" A_i, $i = 1, 2$. Each of these two pieces lies in a component V_i, $i = 1, 2$, of $\overline{\Omega} \setminus (C_2 \cup \gamma)$, respectively. Note that Θ_1, Θ_3 are disjoint from $C_2 \cup \gamma$, since $C_2 \subset \tilde{u}^{-1}(t)$ and $t \neq 0, 1$. Hence, Θ_1 and Θ_3 lie in components of $\overline{\Omega} \setminus (C_2 \cup \gamma)$. We claim that one of the components V_1 and V_2, say V_1, contains neither Θ_1 nor Θ_3. This will be the desired component V with the claimed properties in the previous paragraph. In particular, the arc β is an arc contained in $\partial A_1 \cap \partial Q_i$.

To prove our latter claim, we first observe that the components of the set $\overline{\Omega} \setminus (C_2 \cup \gamma)$, which are rel. open in $\overline{\Omega}$, are pathwise connected and rel. open in $\overline{\Omega}$. This is because $\overline{\Omega}$ is a Jordan region and thus it is locally pathwise connected; see [46, Theorem 27.5 and Theorem 27.9]. The components V_1 and V_2 that contain A_1 and A_2, respectively, are necessarily distinct. Otherwise, there is a path in $\overline{\Omega} \setminus (C_2 \cup \gamma)$ that connects a point $a_1 \in A_1$ to a point $a_2 \in A_2$. Concatenating this path with a path inside Q_i that connects a_1 to a_2 would provide a loop in $\overline{\Omega} \setminus C_2$ that separates C_2, a contradiction to the connectedness of C_2; for the construction of that loop it is crucial that $C_2 \cap Q_i = \emptyset$. Suppose now that V_1 contains Θ_1 and V_2 contains Θ_3. Then we can similarly construct path from Θ_1 to Θ_3 passing through Q_i that disconnects C_2, a contradiction.

Step 2 The points x_2 and x_4 are the only points lying in $u^{-1}(t) \cap \partial Q_i$.

Now, we show that there can be no third point in $u^{-1}(t) \cap \partial Q_i$. In the case $x_2 \neq x_4$, we join the points x_2, x_4 with an arc inside Q_i and we obtain a continuum C that separates Θ_1 and Θ_3, and intersects ∂Q_i in two points. If $x_2 = x_4$ we just let $C = C_2 \cup C_4$. The set $\overline{\Omega} \setminus C$ has at least two components, one containing Θ_1 and one containing Θ_3. If V is one of the components of $\overline{\Omega} \setminus C$, then we have $u \equiv t$ on $\partial_* V$, which implies that $u \geq t$ or $u \leq t$ on $V \cap S$ by the maximum principle in Theorem 3.4.6.

Assume that there exists another point $x \in \partial Q_i \cap u^{-1}(t)$, $x \neq x_2, x_4$. The point x lies on an open arc $\beta \subset \partial Q_i$ with endpoints x_2, x_4. This arc lies in one of the components of $\overline{\Omega} \setminus C$, so assume it lies in a component V on which $u \leq t$. However, by the choice of t, arbitrarily close to x we can find a point $x_+ \in \beta$ with $u(x_+) > t$, a contradiction. $\square$

Proof of Lemma 3.5.7 The set of local maximum values of $f : \mathbb{R} \to \mathbb{R}$ is the set

$$E = \{y \in \mathbb{R} : \text{there exist } x \in \mathbb{R} \text{ and } \varepsilon > 0 \text{ such that}$$
$$y = f(x) \text{ and } f(z) \leq y \text{ for all } |z - x| < \varepsilon\}.$$

We will show that this set is at most countable. The claim for the local minimum values follows by looking at $-f$. Note that $E = \bigcup_{n=1}^{\infty} E_n$, where

$$E_n = \{y \in \mathbb{R} : \text{there exists } x \in \mathbb{R} \text{ such that}$$
$$y = f(x) \text{ and } f(z) \leq y \text{ for all } |z - x| < 1/n\}.$$

Hence, it suffices to show that E_n is at most countable for each $n \in \mathbb{N}$. For each $y \in E_n$ there exists $x \in \mathbb{R}$ such that $f(x) = y$, and there exists an interval $I = (x - 1/n, x + 1/n)$ such that $f(z) \leq y$ for all $z \in I$. If $y_1, y_2 \in E_n$ are distinct with $y_1 = f(x_1)$, $y_2 = f(x_2)$, and I_1, I_2 are the corresponding intervals, then $\frac{1}{2} I_1 := (x_1 - 1/2n, x_1 + 1/2n)$ and $\frac{1}{2} I_2 := (x_2 - 1/2n, x_2 + 1/2n)$ are necessarily disjoint intervals. This implies that E_n is in one-to-one correspondence with a family of disjoint open subintervals of $\mathbb{R}$, and hence E_n is at most countable. $\square$

The next corollary is immediate:

Corollary 3.5.8 *For each peripheral disk Q_i, $i \in \mathbb{N}$, and for a.e. level $t \in [m_{Q_i}, M_{Q_i}]$ the intersection $u^{-1}(t) \cap \partial Q_i$ contains precisely two points.*

Proof Assume that $m_{Q_i}(u) < M_{Q_i}(u)$ (i.e., $\mathrm{osc}_{Q_i}(u) \neq 0$), and choose a $t \in (m_{Q_i}(u), M_{Q_i}(u))$ so that the conclusion of Lemma 3.5.6 is true. Consider two points $p_i, q_i \in \partial Q_i$ such that $u(p_i) = m_{Q_i}(u)$ and $u(q_i) = M_{Q_i}(u)$. Applying the intermediate value theorem on each of the two arcs between the points p_i, q_i, it follows that $u^{-1}(t) \cap \partial Q_i$ contains at least two points. □

Remark 3.5.9 It is clear from the proof of Lemma 3.5.6 that we only need to exclude at most countably many $t \in [m_{Q_i}(u), M_{Q_i}(u)]$ for the conclusion of Corollary 3.5.8.

We continue with an absolute continuity lemma. This is the most technical part of the section. We first observe the following consequence of the Ahlfors 2-regularity of the peripheral disks:

Remark 3.5.10 If a peripheral disk Q_i, $i \in \mathbb{N}$, intersects two circles $\partial B(x, r)$ and $\partial B(x, R)$ with $0 < r < R$, then

$$\mathcal{H}^2(Q_i \cap (B(x, R) \setminus B(x, r))) \geq C(R - r)^2,$$

where $C > 0$ is a constant depending only on the Ahlfors regularity constant of condition (3.2). To see this, by the connectedness of Q_i there exists a point $y \in Q_i \cap \partial B(x, (r + R)/2)$. Then $B(y, (R - r)/2) \subset B(x, R) \setminus B(x, r)$, so

$$\mathcal{H}^2(Q_i \cap (B(x, R) \setminus B(x, r))) \geq \mathcal{H}^2(Q_i \cap B(y, (R - r)/2)) \geq K_1 \frac{(R - r)^2}{4},$$

by the Ahlfors regularity condition (3.2).

Lemma 3.5.11 *For a.e. $t \in [0, 1]$ we have $\mathcal{H}^1(u^{-1}(t)) = 0$.*

The proof is very technical so we provide first a rough sketch of the argument. For a fixed $\varepsilon > 0$ we will find a cover of $S = \bigcup_{t \in [0,1]} u^{-1}(t)$ by balls B_j of radius $r_j < \varepsilon$, so that the total area of the balls is small. We call this cover an effective cover. Then for each t the quantity $\mathcal{H}^1_\varepsilon(u^{-1}(t))$ is bounded by $\sum_{j \in J_t} 2r_j$, where the sum is over the balls intersecting $u^{-1}(t)$. We wish to show that $\mathcal{H}^1_\varepsilon(u^{-1}(t))$ converges to 0 as $\varepsilon \to 0$ for a.e. $t \in [0, 1]$. We prove this by integrating $\sum_{j \in J_t} 2r_j$ over $t \in [0, 1]$, and then showing that the integral converges to 0 as $\varepsilon \to 0$.

Upon integrating, one obtains an expression of the form

$$\sum_j r_j \operatorname{diam}(u(B_j \cap S)),$$

so we wish to find good bounds for $\operatorname{diam}(u(B_j \cap S))$. Thus, we produce bounds using the upper gradient inequality, in combination with the maximum principle (see Lemma 3.4.7):

$$\operatorname{diam}(u(B_j \cap S)) \leq \sum_{i : Q_i \cap \partial B_j \neq \emptyset} \operatorname*{osc}_{Q_i}(u).$$

This is where technicalities arise, because the right hand side is not a good enough bound for all balls B_j.

The bound turns out to be good, in the case the ball B_j intersects only "small" peripheral disks Q_i of diameter $\lesssim r_j$, or in the case the "large" peripheral disks that are possibly intersected by B_j do not have serious contribution to the upper gradient inequality and can be essentially ignored:

$$\operatorname{diam}(u(B_j \cap S)) \lesssim \sum_{\substack{i:Q_i \cap \partial B_j \neq \emptyset \\ Q_i \text{ "small"}}} \underset{Q_i}{\operatorname{osc}}(u).$$

We call "good" the balls B_j satisfying the above.

However, there is a "bad" subcollection of the balls B_j for which the above estimate fails. Namely, these are the balls that intersect some relatively large peripheral disk Q_i, where the latter also has a serious contribution to the upper gradient inequality and cannot be ignored. We amend this by essentially discarding these "bad" balls B_j from our effective cover of the set $\bigcup_{t \in [0,1]} u^{-1}(t)$, and then replacing each of them (in the cover) with a corresponding "large" peripheral disk Q_i (after enlarging it slightly so that we still obtain a cover).

Then, $\mathcal{H}^1_\varepsilon(u^{-1}(t))$ is bounded by

$$\sum 2r_j + \sum \operatorname{diam}(Q_i),$$

where the first sum is over the "good" balls intersecting $u^{-1}(t)$ and the second sum is over the "large" peripheral disks Q_i corresponding to "bad" balls that intersect $u^{-1}(t)$. One now integrates over $t \in [0, 1]$ as before, and treats separately the terms corresponding to the "good" and "bad" balls. We proceed with the details.

Proof By Lemma 3.5.6, for a.e. $t \in (0, 1)$ the level set $u^{-1}(t)$ intersects $\partial \Omega$ and ∂Q_i in at most two points, for all $i \in \mathbb{N}$. Hence, it suffices to show that for a.e. $t \in (0, 1)$ the set $u^{-1}(t) \cap S^\circ$ has $\mathcal{H}^1$-measure equal to zero. Recall that S° contains the points of the carpet not lying on any peripheral circle.

For a fixed $\varepsilon > 0$ consider the finite set $E = \{i \in \mathbb{N} : \operatorname{diam}(Q_i) > \varepsilon\}$. We cover $\Omega \setminus \bigcup_{i \in E} \overline{Q}_i$ by balls B_j of radius $r_j < \varepsilon$ such that $2B_j \subset \Omega \setminus \bigcup_{i \in E} \overline{Q}_i$ and such that $\frac{1}{5} B_j$ are disjoint. The existence of this collection of balls is justified by a basic covering lemma; see e.g. [21, Theorem 1.2].

Let J be the family of indices j such that for each $s \in [1, 2]$ we have

$$\operatorname{diam}(u(sB_j \cap S)) \geq k \underset{Q_i}{\operatorname{osc}}(u) \tag{3.14}$$

for all peripheral disks Q_i with $\operatorname{diam}(Q_i) > 8r_j$ that intersect $\partial(sB_j)$, where $k \geq 1$ is a constant to be determined. It follows from Remark 3.5.10 that for each $j \in J$ there can be at most N_0 such peripheral disks Q_i, where N_0 depends only on the Ahlfors 2-regularity constant. Indeed, each such Q_i must intersect both $\partial(2B_j)$ and $\partial(4B_j)$, so it follows that

$$\mathcal{H}^2(Q_i \cap 4B_j) \geq \mathcal{H}^2(Q_i \cap (4B_j \setminus 2B_j)) \geq Cr_j^2$$

for a uniform constant $C > 0$, depending only on the Ahlfors 2-regularity constant. Comparing the area of $4B_j$ with $\mathcal{H}^2(Q_i \cap 4B_j)$ we arrive at the conclusion.

We fix a ball B_j, $j \in J$, and peripheral disks $Q_{i_1}, \ldots, Q_{i_N}$ as above, where $N \leq N_0$. These are the peripheral disks with diameter bigger than $8r_j$, each of which intersects $\partial(sB_j)$ for some $s \in [1, 2]$ and satisfies (3.14). Our goal is to show that there exists a uniform constant $C > 0$, such that for a.e. $s \in (1, 2)$ we have

$$\text{diam}(u(sB_j \cap S)) \leq C \sum_{\substack{i:Q_i \cap \partial(sB_j) \neq \emptyset \\ i \neq i_1, \ldots, i_N}} \underset{Q_i}{\text{osc}}(u), \qquad (3.15)$$

provided that we choose k suitably, depending only on the data. In other words, the contribution of $\text{osc}_{Q_{i_1}}(u), \ldots, \text{osc}_{Q_{i_N}}(u)$ in the upper gradient inequality is negligible. We fix $s \in (1, 2)$ such that the conclusion of Lemma 3.4.7 holds, i.e.,

$$\text{diam}(u(sB_j \cap S)) \leq \sum_{i:Q_i \cap \partial(sB_j) \neq \emptyset} \underset{Q_i}{\text{osc}}(u).$$

If none of $Q_{i_1}, \ldots, Q_{i_N}$ intersects $\partial(sB_j)$, then (3.15) follows immediately, so we assume that this is not the case. After reordering, suppose that there are M peripheral disks $Q_{i_1}, \ldots, Q_{i_M}$, $M \leq N$, intersecting $\partial(sB_j)$, among $Q_{i_1}, \ldots, Q_{i_N}$. We have

$$\text{diam}(u(sB_j \cap S)) \leq \sum_{\substack{i:Q_i \cap \partial(sB_j) \neq \emptyset \\ i \neq i_1, \ldots, i_M}} \underset{Q_i}{\text{osc}}(u) + \sum_{l=1}^{M} \underset{Q_{i_l}}{\text{osc}}(u)$$

$$\leq \sum_{\substack{i:Q_i \cap \partial(sB_j) \neq \emptyset \\ i \neq i_1, \ldots, i_M}} \underset{Q_i}{\text{osc}}(u) + \frac{M}{k} \text{diam}(u(sB_j \cap S)).$$

We consider $k = 2N_0 \geq 2N \geq 2M$. Then

$$\text{diam}(u(sB_j \cap S)) \leq 2 \sum_{\substack{i:Q_i \cap \partial(sB_j) \neq \emptyset \\ i \neq i_1, \ldots, i_M}} \underset{Q_i}{\text{osc}}(u)$$

and this completes the proof of (3.15).

If we write $B_j = B(x_j, r_j)$, then (3.15) implies that

$$\text{diam}(u(B_j \cap S)) \leq \text{diam}(u(sB_j \cap S)) \leq C \sum_{\substack{i:Q_i \cap \partial B(x_j,s) \neq \emptyset \\ i \neq i_1, \ldots, i_N}} \underset{Q_i}{\text{osc}}(u)$$

for a.e. $s \in (r_j, 2r_j)$. Integrating over $s \in (r_j, 2r_j)$ and applying Fubini's theorem yields

$$r_j \operatorname{diam}(u(B_j \cap S)) \leq C \sum_{\substack{i: Q_i \cap 2B_j \neq \emptyset \\ i \neq i_1, \dots, i_N}} \operatorname*{osc}_{Q_i}(u) \int_{r_j}^{2r_j} \chi_{Q_i \cap \partial B(x_j, s)} \, ds$$

$$\leq C \sum_{\substack{i: Q_i \cap 2B_j \neq \emptyset \\ i \neq i_1, \dots, i_N}} \operatorname*{osc}_{Q_i}(u) \operatorname{diam}(Q_i).$$

We note that if $Q_i \cap 2B_j \neq \emptyset$ and $i \neq i_1, \dots, i_N$, then $\operatorname{diam}(Q_i) \leq 8r_j$ (by the definition of $i_1, \dots, i_N$), so $Q_i \subset 11B_j$. Therefore,

$$r_j \operatorname{diam}(u(B_j \cap S)) \leq C \sum_{i: Q_i \subset 11B_j} \operatorname*{osc}_{Q_i}(u) \operatorname{diam}(Q_i). \tag{3.16}$$

For each $j \in J$ now consider the smallest interval I_j containing $u(B_j \cap S)$, and define $g_\varepsilon(t) = \sum_{j \in J} 2r_j \chi_{I_j}(t)$, $t \in [0, 1]$.

On the other hand, for each $j \notin J$ there exists $s_j \in [1, 2]$ and there exists a peripheral disk Q_i that intersects $\partial(s_j B_j)$ with $\operatorname{diam}(Q_i) > 8r_j$, but $\operatorname{diam}(u(s_j B_j \cap S)) < k \operatorname{osc}_{Q_i}(u)$. Note that some Q_i might correspond to multiple balls B_j, $j \notin J$. Consider the family $\{Q_i\}_{i \in I}$ of all peripheral disks that correspond to balls B_j, $j \notin J$, and for each $i \in I$ let

$$\tilde{Q}_i = \overline{Q}_i \cup \bigcup \{s_j B_j : Q_i \cap \partial(s_j B_j) \neq \emptyset, \operatorname{diam}(Q_i) > 8r_j,$$

$$\text{and } \operatorname{diam}(u(s_j B_j \cap S)) < k \operatorname*{osc}_{Q_i}(u)\}.$$

It is easy to see that for every $\eta > 0$ there exist $s_{j_1} B_{j_1}, s_{j_2} B_{j_2} \subset \tilde{Q}_i$ such that

$$\operatorname{diam}(u(\tilde{Q}_i \cap S)) \leq \operatorname{diam}(u(\partial Q_i)) + \operatorname{diam}(u(s_{j_1} B_{j_1} \cap S))$$

$$+ \operatorname{diam}(u(s_{j_2} B_{j_2} \cap S)) + \eta$$

$$\leq \operatorname*{osc}_{Q_i}(u) + 2k \operatorname*{osc}_{Q_i}(u) + \eta.$$

Hence,

$$\operatorname{diam}(u(\tilde{Q}_i \cap S)) \leq C \operatorname*{osc}_{Q_i}(u) \tag{3.17}$$

for all $i \in I$, where $C = 1 + 2k$ and depends only on the data. Also, observe that $\operatorname{diam}(\tilde{Q}_i) < 2 \operatorname{diam}(Q_i)$, since $\operatorname{diam}(Q_i) > 8r_j$ whenever $s_j B_j \subset \tilde{Q}_i$ and $s_j \leq 2$. For each $i \in I$ consider the smallest interval I_i that contains $u(\tilde{Q}_i \cap S)$, and define

$b_\varepsilon(t) = \sum_{i \in I} 2 \operatorname{diam}(Q_i) \chi_{I_i}(t)$, $t \in [0, 1]$. We remark that $I \cap E = \emptyset$ since the balls $s_j B_j \subset 2B_j$ do not intersect peripheral disks Q_i, $i \in E$.

For each $t \in [0, 1]$ the set $u^{-1}(t) \cap S^\circ$ is covered by the balls B_j, $j \in J$, and the sets $\widetilde{Q}_i$, $i \in I$. Since $r_j < \varepsilon$ for $j \in J$ and $\operatorname{diam}(Q_i) < \varepsilon$ for $i \in I$, we have

$$\mathcal{H}^1_\varepsilon(u^{-1}(t) \cap S^\circ) \le g_\varepsilon(t) + b_\varepsilon(t).$$

It suffices to show that $g_\varepsilon(t) \to 0$ and $b_\varepsilon(t) \to 0$ for a.e. $t \in [0, 1]$, along some sequence of $\varepsilon \to 0$.

Note first that by (3.17) we have

$$\int_0^1 b_\varepsilon(t)\, dt = 2 \sum_{i \in I} \operatorname{diam}(Q_i) \operatorname{diam}(u(\widetilde{Q}_i \cap S))$$

$$\le 2C \sum_{i \in \mathbb{N} \setminus E} \operatorname{diam}(Q_i) \operatorname*{osc}_{Q_i}(u)$$

$$\le 2C \left(\sum_{i \in \mathbb{N} \setminus E} \operatorname{diam}(Q_i)^2 \right)^{1/2} \left(\sum_{i \in \mathbb{N} \setminus E} \operatorname*{osc}_{Q_i}(u)^2 \right)^{1/2}.$$

The first sum is finite by the quasiroundness assumption (3.1), and the second is also finite since $u \in \mathcal{W}^{1,2}(S)$. As $\varepsilon \to 0$, the set E increases to $\mathbb{N}$, hence the sums converge to zero. This shows that $b_\varepsilon \to 0$ in L^1. In particular $b_\varepsilon(t) \to 0$ a.e., along a subsequence.

Finally, we show the same conclusion for $g_\varepsilon(t)$. By (3.16) we have

$$\int_0^1 g_\varepsilon(t)\, dt = \sum_{j \in J} 2r_j \operatorname{diam}(u(B_j \cap S))$$

$$\le 2C \sum_{j \in J} \sum_{i : Q_i \subset 11 B_j} \operatorname*{osc}_{Q_i}(u) \operatorname{diam}(Q_i). \tag{3.18}$$

We define $h(x) = \sum_{i \in \mathbb{N}} (\operatorname*{osc}_{Q_i}(u) / \operatorname{diam}(Q_i)) \cdot \chi_{Q_i}(x)$. By the quasiroundness assumption we have $\operatorname{diam}(Q_i)^2 \simeq \mathcal{H}^2(Q_i)$ for all $i \in \mathbb{N}$, hence the right hand side in (3.18) can be bounded up to a constant by

$$\sum_{j \in J} \int_{11 B_j} h(x)\, d\mathcal{H}^2(x).$$

If Mh denotes the uncentered Hardy–Littlewood maximal function of h (see [21, Sect. 2], the above is bounded up to a constant by

$$\sum_{j \in J} \int_{\frac{1}{5}B_j} Mh(x)\, d\mathcal{H}^2(x) = \int_{\bigcup_{j \in J} \frac{1}{5}B_j} Mh(x)\, d\mathcal{H}^2(x)$$

$$\leq \int_{\Omega \setminus \bigcup_{i \in E} Q_i} Mh(x)\, d\mathcal{H}^2(x),$$

where we use the fact that the balls $\frac{1}{5}B_j$ are disjoint. We wish to show that the latter converges to 0 as $\varepsilon \to 0$. Then, we will indeed have $\int_0^1 g_\varepsilon(t)\, dt \to 0$ as $\varepsilon \to 0$, as desired.

Since $\mathcal{H}^2(S) = 0$, it follows that $\mathcal{H}^2(\Omega \setminus \bigcup_{i \in E} Q_i) \to 0$ as $\varepsilon \to 0$. Note now that $h \in L^2(\Omega)$, with

$$\int_\Omega h^2(x) d\mathcal{H}^2(x) \simeq \sum_{i \in \mathbb{N}} \operatorname*{osc}_{Q_i}(u)^2 = D(u) < \infty.$$

By the L^2-maximal inequality (see [21, Theorem 2.2]) we have $Mh \in L^2(\Omega) \subset L^1(\Omega)$, and this implies that

$$\int_{\Omega \setminus \bigcup_{i \in E} Q_i} Mh(x)\, d\mathcal{H}^2(x) \to 0$$

as $\varepsilon \to 0$.   □

Remark 3.5.12 The Ahlfors 2-regularity of the peripheral disks was crucial in the preceding argument, and it would be interesting if one could relax the assumption of Ahlfors regularity to e.g. a Hölder domain (see [40] for the definition) assumption on the peripheral disks.

Next, we wish to show that for a.e. $t \in [0, 1]$ the level set $\alpha_t = \tilde{u}^{-1}(t)$ is a simple curve that joins Θ_2 and Θ_4. We include a topological lemma.

Lemma 3.5.13 *Let $C \subset \mathbb{R}^2$ be a locally connected continuum with the following property: it is a minimal continuum that connects two distinct points $a, b \in \mathbb{R}^2$. Then C can be parametrized by a simple curve γ.*

The minimality of C is equivalent to saying that any compact proper subset of C is either disconnected, or it does not contain one of the points a, b.

Proof It is a well-known fact that a connected, locally connected compact metric space (also known as a *Peano space*) is *arcwise connected*, i.e., any two points in the space can be joined by a homeomorphic image of the unit interval; see [46, Theorem 31.2, p. 219]. In our case, there exists a homeomorphic copy $\gamma \subset C$ of the unit interval that connects a and b. The minimality of C implies that $C = \gamma$.   □

We prove the next lemma by showing that for a.e. $t \in (0, 1)$ the level set $\alpha_t = \tilde{u}^{-1}(t)$ satisfies the assumptions of the previous lemma.

Lemma 3.5.14 *For a.e. $t \in (0, 1)$ the following are true. The level set α_t is a simple curve that connects Θ_2 to Θ_4. Moreover, if $\alpha_t \cap \overline{Q}_i \neq \emptyset$ for some $i \in \mathbb{N}$, then $\alpha_t \cap \overline{Q}_i$ is precisely an arc with two distinct endpoints on ∂Q_i.*

For the proof we will use the following lemma that we prove later:

Lemma 3.5.15 *Suppose that a compact connected metric space X is not locally connected. Then there exists an open subset U of X with infinitely many connected components having diameters bounded below away from zero.*

Proof of Lemma 3.5.14 By Lemma 3.5.1, for each $t \in (0, 1)$ the level set α_t is simply connected and connects Θ_2 to Θ_4. We choose a $t \in (0, 1)$ that is not a local maximum or local minimum value of u on $\partial \Omega$ or on any peripheral circle ∂Q_i, $i \in \mathbb{N}$. Note that this implies that $\tilde{u}$ is non-constant in Q_i whenever $\alpha_t \cap Q_i \neq \emptyset$. There are countably many such values t that we exclude; see proof of Lemma 3.5.6. Restricting our choice even further, we assume that t is not a critical value of $\tilde{u}$ on any Q_i, $i \in \mathbb{N}$; a non-constant planar harmonic function has at most countably many critical values. Finally, we suppose that we also have $\mathcal{H}^1(u^{-1}(t)) = 0$, which holds for a.e. $t \in (0, 1)$ by Lemma 3.5.11.

Using an argument similar the proof of Lemma 3.5.6 we show that α_t is a minimal continuum connecting Θ_2 to Θ_4. Consider a minimal continuum $C \subset \alpha_t$ joining Θ_2 and Θ_4. The maximum principle from Theorem 3.4.8 implies that on each of the components of $\overline{\Omega} \setminus C$ we have $\tilde{u} \geq t$ or $\tilde{u} \leq t$. If $C \neq \alpha_t$ then there exists a point $x \in \alpha_t \setminus C$, lying in the interior (rel. $\overline{\Omega}$) of one of these components. If $x \in \overline{Q}_i$ for some $i \in \mathbb{N}$ or $x \in \partial \Omega$, by the choice of t, arbitrarily close to x we can find points $x_+, x_- \in \overline{\Omega}$ with $\tilde{u}(x_+) > t$ and $\tilde{u}(x_-) < t$. This is a contradiction.

If $x \in S^\circ$, then we claim that arbitrarily close to x there exists a point $y \in \alpha_t \cap \overline{Q}_i \setminus C$ for some $i \in \mathbb{N}$. Once we establish this claim, we are led to a contradiction by the previous paragraph. Now we prove our claim, which is proved similarly to Lemma 3.3.7. If it failed, then there would exist a small ball $B(x, \varepsilon)$ such that all points $y \in \alpha_t \cap B(x, \varepsilon)$ lie in S°. Since α_t is connected and it exits $B(x, \varepsilon)$, there exists a continuum $\beta \subset \alpha_t \cap B(x, \varepsilon) \cap S^\circ$ with $\mathrm{diam}(\beta) \geq \varepsilon/2$. Then $\mathcal{H}^1(u^{-1}(t)) = \mathcal{H}^1(\alpha_t \cap S) \geq \mathrm{diam}(\beta) > 0$, a contradiction to the choice of t.

Next, we show that α_t is locally connected. Assume that α_t is not locally connected. Then by Lemma 3.5.15 there exists an open set U and $\varepsilon > 0$ such that $U \cap \alpha_t$ contains infinitely many components C_n, $n \in \mathbb{N}$, of diameter at least ε. By passing to a subsequence, we may assume that the continua $\overline{C}_n$ converge to a continuum C in the Hausdorff sense, with $\mathrm{diam}(C) \geq \varepsilon$. By the continuity of $\tilde{u}$, it follows that $C \subset \alpha_t$. We claim that $C \subset S$, so $C \subset u^{-1}(t)$. Assuming the claim, we obtain a contradiction, since $\varepsilon \leq \mathcal{H}^1(C) \leq \mathcal{H}^1(u^{-1}(t)) = 0$.

If $C \cap Q_i \neq \emptyset$, then by shrinking C and C_n we may assume that $C, C_n \subset\subset Q_i$ for all $n \in \mathbb{N}$. By our choice, t is not a critical value of $\tilde{u}$ on Q_i. Finding a local harmonic conjugate $\tilde{v}$, we see that the pair $G := (\tilde{u}, \tilde{v})$ yields a conformal map on a neighborhood of C. Thus for infinitely many $n \in \mathbb{N}$ the continua C_n have large diameter and lie on the preimage under G of a vertical line segment. This contradicts, e.g., the rectifiability of the preimage of this segment.

Our last assertion in the lemma follows from the fact that α_t is a simple curve and from the maximum principle in Theorem 3.4.6. Indeed, by Corollary 3.5.8, for a.e. $t \in [m_{Q_i}(u), M_{Q_i}(u)]$ the intersection $\alpha_t \cap \partial Q_i$ contains precisely two points x, y. If $\alpha_t \cap Q_i = \emptyset$ then α_t connects x, y "externally", and there exists a region $V \subset \Omega$ bounded by α_t and an arc inside Q_i joining x, y. However, by the maximum principle u has to be constant in $V \cap S$. Then $V \cap S \subset \alpha_t$, which contradicts the fact that α_t is a simple curve. Therefore, $\alpha_t \cap Q_i \neq \emptyset$, and since α_t is a simple curve, the conclusion follows.   □

Proof of Lemma 3.5.15 We will use the fact that a space X is locally connected if and only if each component of each open set is open; see [46, Theorem 27.9].

Suppose that X is not locally connected. Then there exists an open set U and a component C_0 of U that is not open. This implies that there exists a point $x \in C_0$ and $\varepsilon > 0$ such that for each $\delta < \varepsilon$ we have $B(x, \delta) \subset \overline{B}(x, \varepsilon) \subset U$ and $B(x, \delta)$ intersects a component C_δ of U, distinct from C_0. We claim that the component C_δ has to meet $\partial B(x, \varepsilon)$, and in particular $\text{diam}(C_\delta) \geq \varepsilon - \delta$. Repeating the argument for a sequence of $\delta \to 0$ yields eventually distinct components C_δ and leads to the conclusion.

Now we prove our claim. Suppose for the sake of contradiction that C_δ does not intersect $\partial B(x, \varepsilon)$. Then C_δ is compact, because it is closed in U (as a component of U) and all of its limit points (in X) are contained in $\overline{B}(x, \varepsilon) \subset U$. Since X is a continuum, there exists a continuum $K \subset X$ that connects C_δ to $\partial B(x, \varepsilon)$ with $K \subset \overline{B}(x, \varepsilon)$; see also the proof of Lemma 3.5.6. In particular, $K \cup C_\delta$ is a connected subset of U, which contradicts the fact that C_δ is a component of U.   □

3.6   The Conjugate Function v

In order to define the conjugate function $v \colon S \to \mathbb{R}$ we introduce some notation. For $i \in \mathbb{N}$, let $\rho(Q_i) = \text{osc}_{Q_i}(u)$, and for a path $\gamma \subset \overline{\Omega}$ define

$$\ell_\rho(\gamma) = \sum_{i : Q_i \cap \gamma \neq \emptyset} \rho(Q_i).$$

We would like to emphasize that we have *not* defined $\text{osc}_{Q_0}(u)$ (where $Q_0 = \mathbb{C} \setminus \overline{\Omega}$), and also terms corresponding to $i = 0$ are not included in the summations of the above type. We will first define a coarse version $\hat{v}$ of the conjugate function that is only defined on the set $\{Q_i\}_{i \in \mathbb{N}}$ of peripheral disks, and then we will define the conjugate function $v(x)$ by taking limits of $\hat{v}(Q_i)$ as Q_i approaches a point $x \in S$. The developments in this section follow the general strategy of [33, Sect. 7]. The definition of the conjugate function here seems to be neater than the one in [33], where there are many technicalities. It would be interesting if the arguments in [33] can be simplified with a more elegant definition of the harmonic conjugate function.

Let $\mathcal{T}$ be the family of $t \in (0, 1)$ for which the conclusions of Theorem 3.5.2, Lemma 3.5.6, Corollary 3.5.8, Lemmas 3.5.11 and 3.5.14 hold. Furthermore, we assume that Lemma 3.5.4 can be applied for the sequence $h(Q_i) = \rho(Q_i)^2$. Finally, we assume that each $t \in \mathcal{T}$ is not a local extremal value of u on $\partial \Omega$ or on any peripheral circle ∂Q_i, $i \in \mathbb{N}$. This, in particular, implies that $\rho(Q_i) > 0$ whenever $\alpha_t \cap Q_i \neq \emptyset$, since otherwise u is constant on ∂Q_i. It also implies that if α_t intersects Q_i, $i \in \mathbb{N}$, then for all sufficiently small $h > 0$ the level set $\alpha_{t \pm h}$ intersects Q_i.

If $\rho(Q_i) > 0$, then for $t \in \mathcal{T} \cap [m_{Q_i}(u), M_{Q_i}(u)]$ the path $\alpha_t = \tilde{u}^{-1}(t)$ intersects ∂Q_i at two points. If we regard the path α_t as a map from $[0, 1]$ into $\overline{\Omega}$, we can parametrize it so that $\alpha_t(0) \in \Theta_2$ and $\alpha_t(1) \in \Theta_4$. We consider an open subpath α_t^i of α_t, terminated at the first entry point of α_t in ∂Q_i. This is to say that α_t^i starts at Θ_2 and terminates at ∂Q_i, while $\alpha_t^i \cap \partial Q_i = \emptyset$. We then define

$$\hat{v}(Q_i) = \inf\{\ell_\rho(\alpha_t^i) : t \in \mathcal{T} \cap [m_{Q_i}(u), M_{Q_i}(u)]\}.$$

Note that we define $\hat{v}(Q_i)$ whenever $\rho(Q_i) > 0$. By Theorem 3.5.2 we have $0 \leq \hat{v}(Q_i) \leq D(u)$ for all $i \in \mathbb{N}$. In fact, the infimum is not needed:

Lemma 3.6.1 *Fix a peripheral disk Q_{i_0}, $i_0 \in \mathbb{N}$. If $s, t \in \mathcal{T} \cap [m_{Q_{i_0}}(u), M_{Q_{i_0}}(u)]$, then $\ell_\rho(\alpha_s^{i_0}) = \ell_\rho(\alpha_t^{i_0})$. In particular, we have $\hat{v}(Q_{i_0}) = \ell_\rho(\alpha_s^{i_0})$.*

Proof Suppose that $s \neq t$. For simplicity, denote $\gamma_s = \alpha_s^{i_0}$ and define $\tilde{\gamma}_t$ to be the smallest open subpath of α_t that connects ∂Q_{i_0} to Θ_4. By Lemma 3.5.14 the curves $\alpha_t^{i_0}$ and $\tilde{\gamma}_t$ meet disjoint sets of peripheral disks. Thus, by Theorem 3.5.2

$$\ell_\rho(\alpha_t^{i_0}) + \rho(Q_{i_0}) + \ell_\rho(\tilde{\gamma}_t) = \ell_\rho(\alpha_t) = D(u). \tag{3.19}$$

For small $h > 0$ so that $s \pm h \in \mathcal{T}$, each of the disjoint curves $\alpha_{s+h}, \alpha_{s-h}$ intersects ∂Q_{i_0} at two points. The strip $\overline{A}_{s-h,s+h}$ is a (closed) Jordan region bounded by the curves $\alpha_{s-h}, \alpha_{s+h}$, and subarcs of Θ_2 and Θ_4; see Proposition 3.5.1 and Lemma 3.5.14. Also, this Jordan region contains $\alpha_s \supset \gamma_s$. Since ∂Q_{i_0} meets both boundary curves α_{s-h} and α_{s+h}, it follows that $\overline{A}_{s-h,s+h} \setminus Q_{i_0}$ has two components, which are (closed) Jordan regions, one intersecting Θ_2 and one intersecting Θ_4. Let $\Omega_{s,h} \subset \overline{A}_{s-h,s+h} \setminus Q_{i_0}$ be the (closed) Jordan region intersecting Θ_2, so $\gamma_s \subset \Omega_{s,h}$. In a completely analogous way we define a closed Jordan region $\tilde{\Omega}_{t,h}$ that intersects Θ_4 and contains $\tilde{\gamma}_t$. Here, we first need to refine our choice of $h > 0$ so that we also have $t \pm h \in \mathcal{T}$. Finally, since $s \neq t$, if $h > 0$ is chosen to be sufficiently small, we may have that the strips $\overline{A}_{s-h,s+h}$ and $\overline{A}_{t-h,t+h}$ are disjoint, and, hence, so are the regions $\Omega_{s,h}$ and $\tilde{\Omega}_{t,h}$.

Following the notation in the proof of Theorem 3.5.3, we define $F_{s,h}$ to be the family of indices $i \in \mathbb{N} \setminus \{i_0\}$ such that $Q_i \cap \alpha_{s-h} \neq \emptyset$, $Q_i \cap \alpha_{s+h} \neq \emptyset$, and $Q_i \cap \gamma_s \neq \emptyset$. In an analogous way we define $F_{t,h}$ that corresponds to $\tilde{\gamma}_t$. Then as

$h \to 0$ we have

$$\sum_{i \in F_{s,h}} \rho(Q_i) \to \sum_{i:Q_i \cap \gamma_s \neq \emptyset} \rho(Q_i) \quad \text{and} \quad \sum_{i \in F_{t,h}} \rho(Q_i) \to \sum_{i:Q_i \cap \tilde{\gamma}_t \neq \emptyset} \rho(Q_i). \quad (3.20)$$

Also, we define $N_h = \{i \in \mathbb{N} : Q_i \cap (A_{s-h,s+h} \cup A_{t-h,t+h}) \neq \emptyset\} \setminus (F_{s,h} \cup F_{t,h} \cup \{i_0\})$.

The set $\overline{\Omega} \setminus (\Omega_{s,h} \cup \tilde{\Omega}_{t,h} \cup Q_{i_0})$ has precisely a component V_1 containing Θ_1 and a component V_3 containing Θ_3. Now, consider the function

$$g(x) = \begin{cases} 0, & x \in V_1 \cap S \\ \frac{u(x)-(s-h)}{2h}, & x \in \Omega_{s,h} \cap S \\ \frac{u(x)-(t-h)}{2h}, & x \in \tilde{\Omega}_{t,h} \cap S \\ 1, & x \in V_3 \cap S. \end{cases}$$

Using Lemma 3.4.4, one can show that $g \in \mathcal{W}^{1,2}(S)$. We provide some details. Let $U_s = \frac{u(x)-(s-h)}{2h} \vee 0$, and consider a bump function $\phi_s : \mathbb{C} \to [0,1]$ that is identically equal to 1 on $\Omega_{s,h}$, and vanishes on $\tilde{\Omega}_{t,h}$. Similarly consider U_t and ϕ_t, which is a bump function equal to 1 on $\tilde{\Omega}_{t,h}$ but vanishes on $\Omega_{s,h}$. Then $U_s \phi_s + U_t \phi_t$ lies in $\mathcal{W}^{1,2}(S)$. We now have

$$g(x) = \begin{cases} U_s(x)\phi_s(x) + U_t(x)\phi(x), & x \in S \setminus V_3 \\ 1, & x \in V_3 \cap S. \end{cases}$$

On $\partial V_3 \cap S$ the two alternatives agree, so by Lemma 3.4.4 we conclude that $g \in \mathcal{W}^{1,2}(S)$.

Also, we have

$$\operatorname*{osc}_{Q_i}(g) \begin{cases} = 1, & i \in F_{s,h} \cup F_{t,h} \cup \{i_0\} \\ \leq \rho(Q_i)/2h, & i \in N_h \\ = 0, & i \notin F_{s,h} \cup F_{t,h} \cup \{i_0\} \cup N_h. \end{cases}$$

We only need to justify the middle inequality. If Q_i, $i \in N_h$, does not intersect $\Omega_{s,h}$ or $\tilde{\Omega}_{t,h}$, then $\operatorname{osc}_{Q_i}(g) = 0$. If Q_i, $i \in N_h$, intersects only one of $\Omega_{s,h}$, $\tilde{\Omega}_{t,h}$, then the desired inequality is clear by the definition of g. If Q_i intersects both $\Omega_{s,h}$ and $\tilde{\Omega}_{t,h}$ then Q_i has to meet either V_1 or V_3. Suppose that it meets V_1. Then the minimum of g on ∂Q_i is 0, and the maximum is attained at a point of $\Omega_{s,h}$ or $\tilde{\Omega}_{t,h}$. Suppose that the maximum is attained at a point of $\Omega_{s,h}$. It follows that both the minimum and the maximum of g on ∂Q_i are attained in $\Omega_{s,h}$, and the definition of g implies that the oscillation is bounded by $\rho(Q_i)/2h$, as desired. The other cases yield the same conclusion.

Since every convex combination of u and g is admissible for the free boundary problem, as in the proof of Proposition 3.5.3 (see (3.10)), it follows that

$$D(u) \leq \sum_{i \in \mathbb{N}} \rho(Q_i) \operatorname*{osc}_{Q_i}(g).$$

Thus,

$$D(u) \leq \sum_{i \in F_{s,h}} \rho(Q_i) + \sum_{i \in F_{t,h}} \rho(Q_i) + \rho(Q_{i_0}) + \frac{1}{2h} \sum_{i \in N_h} \rho(Q_i)^2.$$

Letting $h \to 0$ and using (3.20), together with Lemma 3.5.4, we obtain

$$D(u) \leq \ell_\rho(\gamma_s) + \ell_\rho(\tilde{\gamma}_t) + \rho(Q_{i_0}) = \ell_\rho(\alpha_s^{i_0}) + \ell_\rho(\tilde{\gamma}_t) + \rho(Q_{i_0}).$$

By (3.19) we obtain $\ell_\rho(\alpha_t^{i_0}) \leq \ell_\rho(\alpha_s^{i_0})$. The roles of s, t is symmetric, so the conclusion follows. □

If a point $x \in S$ has arbitrarily small neighborhoods that contain some Q_i with $\rho(Q_i) > 0$, then we define

$$v(x) = \liminf_{Q_i \to x, x \notin Q_i} \hat{v}(Q_i), \qquad (3.21)$$

where in the limit we only consider peripheral disks for which $\rho(Q_i) > 0$, since for the other peripheral disks $\hat{v}$ is not defined. Observe that $0 \leq v(x) \leq D(u) < \infty$. If $x \in \alpha_t \cap S$ for some $t \in \mathcal{T}$, then we can approximate x by peripheral disks Q_i that intersect α_t. This follows from Lemma 3.3.7 because $\mathcal{H}^1(\alpha_t \cap S) = 0$ by Lemma 3.5.11. All these peripheral disks have $\rho(Q_i) > 0$ and thus $v(x)$ can be defined by the preceding formula.

If $v(x_0)$ cannot be defined, this means that there exists a neighborhood of x_0 that *contains* only peripheral disks with $\rho(Q_i) = 0$; note that x_0 could lie on some peripheral circle ∂Q_{i_0} with $\rho(Q_{i_0}) > 0$. The continuity of u and the upper gradient inequality imply that u is constant in this neighborhood (see Lemma 2.7.2), and in particular it takes some value t_0, where $t_0 \notin \mathcal{T}$. We define $U = \mathrm{int}_{\overline{\Omega}}(\alpha_{t_0})$ (i.e., the interior of α_{t_0} rel. $\overline{\Omega}$) and observe that if x_0 does not lie on a peripheral circle ∂Q_{i_0} with $\rho(Q_{i_0}) > 0$, then $x_0 \in U$. In particular, this is true if $x_0 \in S^\circ$. If x_0 lies on some peripheral circle with $\rho(Q_{i_0}) > 0$, then $x_0 \in \overline{V}$, where V is a component of U. The set V is rel. open in $\overline{\Omega}$ because components of an open subset of a locally connected space are open (see [46, Theorem 27.9]). Furthermore, if V intersects a peripheral disk Q_i, then $\rho(Q_i) = 0$ and $Q_i \subset V$. Indeed, if $V \cap Q_i \neq \emptyset$ then the (classical) harmonic function $\tilde{u}|_{Q_i}$ is constant, equal to t_0, since it is constant on an open set. Another observation is that V is pathwise connected. This is because it is a component of the open subset U of $\overline{\Omega}$, and $\overline{\Omega}$ is locally pathwise connected; see [46, Theorem 27.5 and Theorem 27.9]. In fact, $V \cap \Omega$ is pathwise connected, because any path in V that connects two points of $V \cap \Omega$ is homotopic to a path in $V \cap \Omega$ that connects the same points.

The following lemma allows us to define v on all of S.

Lemma 3.6.2 *Let V be a component of $\mathrm{int}_{\overline{\Omega}}(\alpha_{t_0})$ for some $t_0 \notin \mathcal{T}$. Then the function v has the same value on the points $x \in \partial_{\overline{\Omega}} V \cap S$ for which the formula (3.21) is applicable, and there exists at least one such point.*

Here $\partial_{\overline{\Omega}} V$ is the boundary of V rel. $\overline{\Omega}$. Hence, the lemma allows us to define v to be constant on $\overline{V} \cap S$.

For this proof and other proofs in this section we will use repeatedly Lemma 3.13.1 from Sect. 3.13.

Proof We will split the proof in four cases. We give details in the proof of Case 0 below, and for the rest of the cases we will describe the variational argument that has to be used, skipping some of the details.

Case 0 V contains only one peripheral disk Q_{i_0}. In this case, $V = Q_{i_0}$. This is because V is rel. open in $\overline{\Omega}$ and if it contained points outside Q_{i_0} it would also contain other peripheral disks.

By the discussion preceding the statement of the lemma, it follows that on every point of ∂Q_{i_0} the formula (3.21) is applicable. Also, $\tilde{u} \equiv t_0$ on Q_{i_0}, thus $\rho(Q_{i_0}) = 0$. By a variational argument we now show that v is constant on ∂Q_{i_0}. Let $a, b \in \partial Q_{i_0}$ be arbitrary points, and assume there exists $\varepsilon > 0$ such that $v(b) - v(a) \geq 10\varepsilon$. Using Lemma 3.13.1, for every $\eta > 0$ we can find a test function $\zeta \in \mathcal{W}^{1,2}(S)$ that vanishes on $\partial\Omega$ with $0 \leq \zeta \leq 1$, such that $\zeta \equiv 1$ on small disjoint balls $B(a, r) \cup B(b, r) \subset \Omega$, and

$$D(\zeta) - \mathrm{osc}_{Q_{i_0}}(\zeta)^2 < \eta. \tag{3.22}$$

Using (3.21), we can find peripheral disks $Q_{i_a} \subset B(a, r)$ and $Q_{i_b} \subset B(b, r)$ with $\rho(Q_{i_a}) > 0$, $\rho(Q_{i_b}) > 0$ and

$$\hat{v}(Q_{i_b}) - \hat{v}(Q_{i_a}) > 9\varepsilon. \tag{3.23}$$

By Lemma 3.6.1, we can consider $s, t \in \mathcal{T}$, $s \neq t$, such that for the smallest open subpaths γ_s, γ_t of α_s, α_t, respectively, that connect Θ_2 to Q_{i_a}, Q_{i_b}, respectively, we have

$$\hat{v}(Q_{i_a}) = \ell_\rho(\gamma_s) \quad \text{and} \quad \hat{v}(Q_{i_b}) = \ell_\rho(\gamma_t). \tag{3.24}$$

We also denote by $\tilde{\gamma}_t$ the smallest open subpath of α_t that connects Q_{i_b} to Θ_4. By Theorem 3.5.2 we have

$$\ell_\rho(\gamma_t) + \ell_\rho(\tilde{\gamma}_t) \leq \ell_\rho(\alpha_t) = D(u).$$

Thus, by (3.24)

$$\hat{v}(Q_{i_a}) - \hat{v}(Q_{i_b}) \geq \ell_\rho(\gamma_s) + \ell_\rho(\tilde{\gamma}_t) - D(u)$$

We claim that

$$\ell_\rho(\gamma_s) + \ell_\rho(\tilde{\gamma}_t) \geq D(u) - \varepsilon. \tag{3.25}$$

This, together with the previous inequality, contradicts (3.23).

We now focus on proving (3.25). As in the proof of Lemma 3.6.1, we can consider a small $h > 0$ and disjoint closed Jordan regions $\Omega_{s,h}$ containing γ_s and $\tilde{\Omega}_{t,h}$ containing $\tilde{\gamma}_t$ such that $\Omega_{s,h}$ connects Q_{i_a} to Θ_2 and $\tilde{\Omega}_{t,h}$ connects Q_{i_b} to Θ_4; see Fig. 3.2. Define $F_{s,h}$ to be the set of indices $i \in \mathbb{N}$ such that $Q_i \cap \gamma_s \neq \emptyset$, $Q_i \cap \alpha_{s-h} \neq \emptyset$, and $Q_i \cap \alpha_{s+h} \neq \emptyset$. Define similarly $F_{t,h}$ that corresponds to $\tilde{\Omega}_{t,h}$, and set $N_h = \{i \in \mathbb{N} : Q_i \cap (A_{s-h,s+h} \cup A_{t-h,t+h}) \neq \emptyset\} \setminus (F_{s,h} \cup F_{t,h})$.

Now, we define carefully an admissible function that lies in $\mathcal{W}^{1,2}(S)$. Let $\phi_s : \overline{\Omega} \to [0, 1]$ be a smooth function that is equal to 1 on $\Omega_{s,h}$ but vanishes outside $\Omega_{s,2h} \cup Q_{i_a}$ (choose a smaller $h > 0$ if necessary so that the Jordan region $\Omega_{s,2h}$ still connects Θ_2 to Q_{i_0}). Similarly, define ϕ_t to be 1 on $\tilde{\Omega}_{t,h}$, and 0 outside $\tilde{\Omega}_{t,2h} \cup Q_{i_b}$.

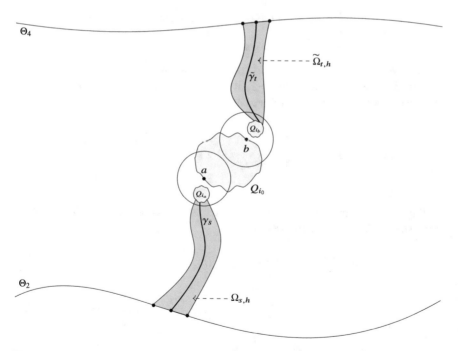

Fig. 3.2 The curves γ_s and $\tilde{\gamma}_t$, and the corresponding Jordan regions $\Omega_{s,h}$ and $\tilde{\Omega}_{t,h}$

Then define

$$U_s(x) = \begin{cases} \frac{u(x)-(s-h)}{h} \vee 0, & u(x) < s \\ \frac{s+h-u(x)}{h} \vee 0, & u(x) \geq s \end{cases} = \left(1 - \frac{|u(x)-s|}{h}\right) \vee 0 \qquad (3.26)$$

for $x \in S$, and similarly define $U_t(x)$ where s is replaced by t in the previous definition. These functions are supported on the sets $\{s - h < u < s + h\} = A_{s-h,s+h} \cap S$ and $\{t - h < u < t + h\} = A_{t-h,t+h} \cap S$, respectively. Also, they lie in the Sobolev space $\mathcal{W}^{1,2}(S)$ by Lemma 3.4.4. We consider their truncation $U_s\phi_s + U_t\phi_t$, and then take $(U_s\phi_s + U_t\phi_t) \vee \zeta$. This function again lies in $\mathcal{W}^{1,2}(S)$, but it vanishes on Θ_1 and Θ_3, so it is not yet admissible for the free boundary problem. To turn it into an admissible function, we would like to set it equal to 1 near Θ_3.

Consider the union of the following paths: γ_s that joins Θ_2 to Q_{i_a}, a line segment in $B(a, r)$ connecting the endpoint of γ_s to a, an arc inside Q_{i_0} connecting a to b, a line segment in $B(b, r)$ connecting b to the endpoint of $\tilde{\gamma}_t$, and $\tilde{\gamma}_t$, where the latter connects Θ_4 to Q_{i_b}. This union separates Θ_1 from Θ_3, so it contains a simple path γ that connects Θ_2 to Θ_4 and separates Θ_1 from Θ_3 (see e.g. the proof of Lemma 3.5.13). Let W be the component of $\overline{\Omega} \setminus \gamma$ that contains Θ_3. We define

$$g(x) = \begin{cases} (U_s(x)\phi_s(x) + U_t(x)\phi_t(x)) \vee \zeta(x), & x \in S \setminus W \\ 1, & x \in W \cap S. \end{cases}$$

On $\gamma \cap S$ the function g is equal to 1 and the two alternatives agree, so by Lemma 3.4.4 we conclude that $g \in \mathcal{W}^{1,2}(S)$ and g is admissible. Furthermore, by construction and Lemma 3.4.4 we have

$$\operatorname*{osc}_{Q_i}(g) \leq \operatorname*{osc}_{Q_i}(\zeta) + \begin{cases} 1, & i \in F_{s,h} \cup F_{t,h} \cup \{i_0\} \\ \rho(Q_i)/2h, & i \in N_h \\ 0, & i \notin F_{s,h} \cup F_{t,h} \cup \{i_0\} \cup N_h. \end{cases}$$

Testing the minimizing property of u against g, as usual (see the proof of Proposition 3.5.3 and (3.10)), we obtain

$$D(u) \leq \sum_{i \in \mathbb{N}} \rho(Q_i) \operatorname*{osc}_{Q_i}(g)$$

$$\leq \sum_{i \in \mathbb{N}} \rho(Q_i) \operatorname*{osc}_{Q_i}(\zeta) + \sum_{i \in F_{s,h}} \rho(Q_i) + \sum_{i \in F_{t,h}} \rho(Q_i)$$

$$+ \rho(Q_{i_0}) + \frac{1}{2h} \sum_{i \in N_h} \rho(Q_i)^2.$$

Letting $h \to 0$ and using Lemma 3.5.4 we obtain

$$
\begin{aligned}
D(u) &\leq \sum_{i \in \mathbb{N} \setminus \{i_0\}} \rho(Q_i) \operatorname*{osc}_{Q_i}(\zeta) + \ell_\rho(\gamma_s) + \ell_\rho(\tilde{\gamma}_t) + \rho(Q_{i_0}) \\
&\leq D(u)^{1/2}(D(\zeta) - \operatorname*{osc}_{Q_{i_0}}(\zeta)^2)^{1/2} + \ell_\rho(\gamma_s) + \ell_\rho(\tilde{\gamma}_t) + \rho(Q_{i_0}) \qquad (3.27) \\
&\leq D(u)^{1/2} \eta^{1/2} + \ell_\rho(\gamma_s) + \ell_\rho(\tilde{\gamma}_t) + \rho(Q_{i_0}),
\end{aligned}
$$

where we used (3.22). Note that $\rho(Q_{i_0}) = 0$. If η is chosen to be sufficiently small, depending on ε, then (3.25) is satisfied and our claim is proved. We have completed the proof of Case 0.

Now, we analyze the remaining cases. If V contains more than one peripheral disk, then it also contains a point $x_0 \in S^\circ$, as one can see by connecting two peripheral disks of V with a path inside V (recall that V is pathwise connected by the discussion preceding Lemma 3.6.2) and then applying Corollary 3.3.6. Using Lemma (b), we can find a path $\beta \subset S^\circ$ that connects x_0 to a side Θ_l, where $l = 1$ or $l = 3$ and Θ_l is such that it is not intersected by $V \subset \alpha_{t_0}$ (recall that $\Theta_1 \subset \alpha_0$ and $\Theta_3 \subset \alpha_1$). Then β has to exit V, so it meets $\partial_{\overline{\Omega}} V \cap S^\circ$ at a point a. Since $a \in S^\circ$, it follows that $v(a)$ is defined by (3.21). Indeed, by the discussion preceding Lemma 3.6.2, if $a \in S^\circ$ and $v(a)$ cannot be defined by (3.21), then we necessarily have $a \in \operatorname{int}_{\overline{\Omega}}(\alpha_{t_0})$ so $a \notin \partial_{\overline{\Omega}} V$. We claim that for any point $b \in \partial_{\overline{\Omega}} V \cap S$ for which definition (3.21) applies we have $v(b) = v(a)$.

Case 1 $b \in S^\circ$; see Fig. 3.3. We keep the same notation as in Case 0. Since $a, b \in S^\circ$, by Lemma 3.13.1, for every $\eta > 0$ we can find a test function ζ that vanishes on $\partial \Omega$ with $0 \leq \zeta \leq 1$, such that $\zeta \equiv 1$ on disjoint balls $B(a, r), B(b, r) \subset \Omega$ and $D(\zeta) < \eta$. We choose $Q_{i_a} \subset B(a, r)$ and $Q_{i_b} \subset B(b, r)$ as before, and heading for a contradiction, we wish to show again (3.25) using a variational argument.

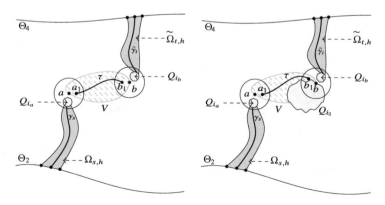

Fig. 3.3 Case 1 (left) and Case 2b (right)

Consider, as in Case 0 the functions U_s, U_t, ϕ_s, ϕ_t with exactly the same definition, and the function $(U_s\phi_s + U_t\phi_t) \vee \zeta$ (we choose a sufficiently small $h > 0$ so that this function vanishes on Θ_1 and Θ_3). Also, consider the union of the following paths: γ_s that joins Θ_2 to Q_{i_a}, a line segment in $B(a, r)$ connecting the endpoint of γ_s to a point $a_1 \in V \cap B(a, r)$, a simple path τ inside the pathwise connected open set $V \cap \Omega$ joining a_1 to a point $b_1 \in V \cap B(b, r)$, a line segment in $B(b, r)$ joining b_1 to the endpoint of $\tilde{\gamma}_t$, and $\tilde{\gamma}_t$. This union contains a simple path γ that separates Θ_1 from Θ_3. However, the function $(U_s\phi_s + U_t\phi_t) \vee \zeta$ need not be equal to 1 on $\gamma \cap \tau \cap S$. We amend this as follows.

Let $\delta > 0$ be so small that $N_\delta(\tau) \subset\subset V \cap \Omega$. Consider the Lipschitz function $\psi(x) = \max\{1 - \delta^{-1} \mathrm{dist}(x, \tau), 0\}$ on $\overline{\Omega}$. Since $\psi \equiv 1$ on τ, it follows that the function $(U_s\phi_s + U_t\phi_t) \vee \zeta \vee \psi$ is equal to 1 on $\gamma \cap S$. Let W be the component of $\overline{\Omega} \setminus \gamma$ containing Θ_3, and define

$$g(x) = \begin{cases} (U_s(x)\phi_s(x) + U_t(x)\phi_t(x)) \vee \zeta(x) \vee \psi(x), & x \in S \setminus W \\ 1, & x \in W \cap S. \end{cases}$$

This function lies in $\mathcal{W}^{1,2}(S)$ by Lemma 3.4.4 and it is admissible. By construction, using notation from Case 0 we have

$$\mathop{\mathrm{osc}}_{Q_i}(g) \leq \mathop{\mathrm{osc}}_{Q_i}(\zeta) + \begin{cases} 1, & i \in F_{s,h} \cup F_{t,h} \\ \rho(Q_i)/2h, & i \in N_h \\ 0, & i \notin F_{s,h} \cup F_{t,h} \cup N_h \cup I_V \\ \delta^{-1}\mathrm{diam}(Q_i), & i \in I_V. \end{cases}$$

Recall that $I_V = \{i \in \mathbb{N} : Q_i \cap V \neq \emptyset\}$ and that $Q_i \subset V$ whenever $Q_i \cap V \neq \emptyset$; see remarks before Lemma 3.6.2. Testing the minimizing property of u against g we obtain

$$D(u) \leq \sum_{i\in\mathbb{N}} \rho(Q_i) \mathop{\mathrm{osc}}_{Q_i}(g)$$

$$\leq \sum_{i\in\mathbb{N}} \rho(Q_i) \mathop{\mathrm{osc}}_{Q_i}(\zeta) + \sum_{i\in F_{s,h}} \rho(Q_i) + \sum_{i\in F_{t,h}} \rho(Q_i) \qquad (3.28)$$

$$+ \frac{1}{2h} \sum_{i\in N_h} \rho(Q_i)^2 + \sum_{i:Q_i\subset V} \delta^{-1}\rho(Q_i)\,\mathrm{diam}(Q_i).$$

Note that $\rho(Q_i) = 0$ for all $Q_i \subset V$, so the last term vanishes identically. By letting $h \to 0$ and choosing a sufficiently small $\eta > 0$, we again arrive at (3.25) and this completes the proof.

Case 2 $b \in \partial Q_{i_0}$ for some $i_0 \in \mathbb{N}$. Since $b \in \partial Q_{i_0}$ (as in Case 0), the difficulty is that for the test function ζ (equal to 1 near b) we have $\mathrm{osc}_{Q_{i_0}}(\zeta) = 1$. Thus, in the variational argument the term $\sum_{i \in \mathbb{N}} \rho(Q_i) \operatorname{osc}_{Q_i}(\zeta)$ is not small, unless $\rho(Q_{i_0}) = 0$, which was true in Case 0. We assume here that $\rho(Q_{i_0}) > 0$, since otherwise the argument is similar, but simpler. Then $Q_{i_0} \cap V = \emptyset$, by the remarks before the statement of Lemma 3.6.2.

As in the other cases, we assume $v(b) - v(a) \geq 10\varepsilon$ and consider a test function $\zeta \in \mathcal{W}^{1,2}(S)$ that vanishes on $\partial\Omega$ with $\zeta \equiv 1$ on $B(a, r) \cup B(b, r)$, and $D(\zeta) - \operatorname{osc}_{Q_{i_0}}(\zeta)^2 < \eta$. Furthermore, by the definition (3.21), we may take $Q_{i_a} \subset B(a, r)$ and $Q_{i_b} \subset B(b, r)$ such that $\hat{v}(Q_{i_a}) - \hat{v}(Q_{i_b}) > 9\varepsilon$, and consider as in Case 0 the open paths γ_s and $\tilde{\gamma}_t$. Again, we are aiming for (3.25). We now split into two sub-cases:

Case 2a $\tilde{\gamma}_t \cap Q_{i_0} \neq \emptyset$. In this case, we use exactly same function $g \in \mathcal{W}^{1,2}(S)$ that we used in Case 1, with the observation that $i_0 \in F_{t,h}$, and $\operatorname{osc}_{Q_{i_0}}(g) \leq 1$. Hence, (for all sufficiently small $h > 0$) we obtain the inequality (3.28) with the first sum replaced by a sum over $i \in \mathbb{N} \setminus \{i_0\}$, because the term $i = i_0$ is included in the third sum. The fact that $\sum_{i \in \mathbb{N} \setminus \{i_0\}} \rho(Q_i) \operatorname{osc}_{Q_i}(\zeta)$ can be made arbitrarily small, leads to the conclusion.

Case 2b $\tilde{\gamma}_t \cap Q_{i_0} = \emptyset$; see Fig. 3.3. In this case, we may assume that $\gamma_s \cap Q_{i_0} = \emptyset$ too, since otherwise we are essentially reduced to the previous case, where the summation index $i = i_0$ is included in the second sum appearing in (3.28).

We construct a path γ that separates Θ_1 from Θ_3 and does not intersect Q_{i_0}. Consider the union of the following continua: γ_s, a line segment in $B(a, r)$ joining the endpoint of γ_s to a point $a_1 \in V \cap B(a, r)$, a simple path $\tau \subset V$ connecting a_1 to a point $b_1 \in V \cap B(b, r)$, a *simple path γ_b in $B(b, r) \setminus Q_{i_0}$* connecting b_1 to the endpoint of $\tilde{\gamma}_t$, and $\tilde{\gamma}_t$. For the existence of γ_b note that if Q_{i_b} is sufficiently close to b then it has to lie in the component of $B(b, r) \setminus Q_{i_0}$ that contains b in its boundary; here one uses the local connectedness property of the Jordan curve ∂Q_{i_0}. Now we let γ be a simple path contained in this union, and separating Θ_1 from Θ_3. Since $Q_{i_0} \cap V = \emptyset$, it follows that $\tau \cap Q_{i_0} = \emptyset$. To ensure that $\gamma \cap Q_{i_0} = \emptyset$, one only has to take a possibly smaller ball $B(a, r)$, so that the line segment we considered there does not meet Q_{i_0}; recall that $a \in S^\circ$.

Let W be the component of $\overline{\Omega} \setminus \gamma$ that contains Θ_3, and by construction $Q_{i_0} \subset W$ or $Q_{i_0} \subset \overline{\Omega} \setminus W$. If $Q_{i_0} \subset W$, then we consider the variation g as in Case 1, which is admissible for a sufficiently small $h > 0$. Since $g = 1$ on ∂Q_{i_0}, we have $\operatorname{osc}_{Q_{i_0}}(g) = 0$. Hence, we may have (3.28) with the first sum replaced by $\sum_{i \in \mathbb{N} \setminus \{i_0\}} \rho(Q_i) \operatorname{osc}_{Q_i}(\zeta)$ and this completes the proof.

Assume now that $Q_{i_0} \subset \overline{\Omega} \setminus W$. Consider the function $g_0 = (U_s\phi_s + U_t\phi_t) \vee \zeta \vee \psi$ as in Case 1, and define $g_1 = 1$ on $S \setminus W$, and $g_1 = g_0$ on $S \cap W$. Then $g_2 := 1 - g_1$ is admissible and satisfies $\operatorname{osc}_{Q_{i_0}}(g_2) = 0$, which yields the result.

Case 3 $b \in \partial\Omega$. The complication here is that our test function ζ does not vanish at $\partial\Omega$. It turns out though that we can always construct an admissible function using the procedure in the Case 2b, as we explain below.

We consider a path γ as before that separates Θ_1 and Θ_3 such that γ contains a simple path $\tau \subset V$ as in Case 2. The function $g_0 = (U_s\phi_s + U_t\phi_t) \vee \zeta \vee \psi$ is equal to 1 on $\gamma \cap S$. Let W be the component of $\overline{\Omega} \setminus \gamma$ that contains Θ_3. If $g_0 = 1$ on points of Θ_3, then $g_0 = 0$ on Θ_1 (by choosing a sufficiently small $h > 0$ and a ζ with small support), so we can set $g = 1$ on $S \cap W$ and $g = g_0$ on $S \setminus W$. If $g_0 = 1$ on points of Θ_1 then we "flip" the function g. We set $g_1 = 1$ on $S \setminus W$ and $g_1 = g_0$ on $S \cap W$. Then $g_2 := 1 - g_1$.

The energy $D(\zeta)$ can be made arbitrarily small because $b \in \partial\Omega$ (recall Lemma 3.13.1), so $\sum_{i \in \mathbb{N}} \rho(Q_i) \operatorname{osc}_{Q_i}(\zeta)$ can be made arbitrarily small in (3.28). In either case, running the variational argument with the admissible function g or g_2 will yield the conclusion. $\square$

Now we can define v on all of S as follows. Let $x \in S$. From the analysis preceding the statement of Lemma 3.6.2 we see that either $v(x)$ can be defined by the formula (3.21), or $x \in \overline{V} \cap S$, where V is a component of $U := \operatorname{int}_{\overline{\Omega}}(\alpha_{t_0})$ for some $t_0 \notin \mathcal{T}$. In the latter case, we define $v(x)$ to be the constant value of v on the points of $\partial_{\overline{\Omega}} V \cap S$ for which (3.21) is applicable; there exists at least one such point by Lemma 3.6.2.

Following a similar argument to the proof of Lemma 3.6.2, we show that v is continuous on S.

Lemma 3.6.3 *The function* $v \colon S \to \mathbb{R}$ *is continuous.*

Proof The proof uses essentially the same variational arguments as in the proof of Lemma 3.6.2, so we skip most of the details. We will also use the same notation as in Lemma 3.6.2. This time we do not need to use the function ψ, and we only need to consider a test function ζ that is supported in one small ball around the point that we wish to show continuity, rather than having two disjoint balls (so the variational arguments here are simpler). By the definition of v we only need to prove continuity for points x_0 such that $v(x_0)$ is defined by (3.21). Indeed, if x_0 cannot be defined by (3.21), then x_0 has a neighborhood in S where v is constant. This is because this neighborhood is contained in $\overline{V}$, where V is a component of some set $U = \operatorname{int}_{\overline{\Omega}}(\alpha_{t_0})$ for a level $t_0 \notin \mathcal{T}$; see also the comments before the statement of Lemma 3.6.2.

For $\eta > 0$ consider a function $\zeta \in \mathcal{W}^{1,2}(S)$ supported around x_0, with $D(\zeta) < \eta$ and $\zeta \equiv 1$ in $B(x_0, r)$. If continuity at x_0 fails, then there exists $\varepsilon > 0$ and a point $y_0 \in B(x_0, r) \cap S$ arbitrarily close to x_0 such that, say, $v(y_0) - v(x_0) > 10\varepsilon$. We claim that there exist peripheral disks $Q_{i_a}, Q_{i_b} \subset B(x_0, r)$ arbitrarily close to x_0 such that $\rho(Q_{i_a}), \rho(Q_{i_b}) > 0$ and

$$\hat{v}(Q_{i_b}) - \hat{v}(Q_{i_a}) > 9\varepsilon. \tag{3.29}$$

Note that by the definition (3.21) there exists a peripheral disk Q_{i_a} arbitrarily close to x_0 with $\rho(Q_{i_a}) > 0$ and $|v(x_0) - \hat{v}(Q_{i_a})| < \varepsilon/2$. If $v(y_0)$ can be defined by (3.21), then we can find a peripheral disk Q_{i_b} arbitrarily close to y_0 (and thus close to x_0) such that $|v(y_0) - \hat{v}(Q_{i_b})| < \varepsilon/2$. We now easily obtain (3.29). If y_0 cannot be defined by (3.21), then $y_0 \in \overline{V}$, where V is a component of $U := \operatorname{int}_{\overline{\Omega}}(\alpha_{t_0})$ for some $t_0 \notin \mathcal{T}$. Note that $x_0 \notin \overline{V}$, otherwise we would have $v(x_0) = v(y_0)$

by Lemma 3.6.2. If y_0 is sufficiently close to x_0, then by Lemma (b)(a) we may consider a path $\gamma \subset B(x, r) \cap S°$ that connects x_0 and y_0. Parametrizing γ as it runs from x_0 to y_0, we consider its first entry point z_0 in $\partial_{\overline{\Omega}} V \cap S°$. Then we necessarily have that $v(y_0) = v(z_0)$ and $v(z_0)$ can be defined by (3.21); see comments before Lemma 3.6.2. Thus, there exists a small Q_{i_b} near z_0 such that $|v(z_0) - \hat{v}(Q_{i_b})| < \varepsilon/2$. If y_0 is very close to x_0 then $\gamma \ni z_0$ is also very close to x_0 (by Lemma (b)) and thus Q_{i_b} is close to x_0. Now (3.29) follows from the above and from the assumption that $v(z_0) - v(x_0) = v(y_0) - v(x_0) > 10\varepsilon$.

By Lemma 3.6.1 we consider $s, t \in \mathcal{T}$ such that $\hat{v}(Q_{i_a}) = \ell_\rho(\gamma_s)$ and $\hat{v}(Q_{i_b}) = \ell_\rho(\gamma_t)$ where γ_s and γ_t are the smallest open subpaths of α_s, α_t that connect Θ_2 to Q_{i_a}, Q_{i_b}, respectively. Also, denote by $\tilde{\gamma}_t$ the smallest open subpath of α_t that connects Q_{i_b} to Θ_4. Arguing as in the beginning of the proof of Lemma 3.6.2, it suffices to show (3.25) to obtain a contradiction. We will split again into cases, and sketch the variational argument that has to be used. Recall that will use the notation from the proof of Lemma 3.6.2.

Case 1 $x_0 \in S°$. We consider the function $g_0 = (U_s \phi_s + U_t \phi_t) \vee \zeta$, and a simple path γ that separates Θ_1 from Θ_3, such that $g_0 = 1$ on $\gamma \cap S$. Let W be the component of $\overline{\Omega} \setminus \gamma$ that contains Θ_3. We define $g = 1$ on $S \cap W$ and $g = g_0$ on $S \setminus W$. This yields an admissible function, as long as $h > 0$ is sufficiently small so that g_0 vanishes on Θ_1. Then by testing the minimizing property of u against g we obtain the conclusion.

Case 2 $x_0 \in \partial Q_{i_0}$ for some $i_0 \in \mathbb{N}$.

Case 2a At least one of the sets $\gamma_s, \tilde{\gamma}_t$ meets Q_{i_0}. This is treated similarly to Case 2a in Lemma 3.6.2. The observation is that we have either $i_0 \in F_{s,h}$ or $i_0 \in F_{t,h}$.

Case 2b None of $\gamma_s, \tilde{\gamma}_t$ meets Q_{i_0}. This is similar to the Case 2b in Lemma 3.6.2, since we can construct a simple path γ that separates Θ_1 from Θ_3 so that the function $g_0 = (U_s \phi_s + U_t \phi_t) \vee \zeta$ is equal to 1 on $\gamma \cap S$ and vanishes on Θ_1 and Θ_3.

Case 3 $x_0 \in \partial\Omega$. We consider a path γ as before such that $g_0 = 1$ on $\gamma \cap S$, and let W be the component of $\overline{\Omega} \setminus \gamma$ that contains Θ_3. If $g_0 = 1$ on points of Θ_3, then $g_0 = 0$ on Θ_1 (by choosing a ζ with small support), so we can set $g = 1$ on $S \cap W$ and $g = g_0$ on $S \setminus W$. If $g_0 - 1$ on points of Θ_1 then we "flip" the function g. We set $g_1 = 1$ on $S \setminus W$ and $g_1 = g_0$ on $S \cap W$. Then $g_2 := 1 - g_1$. In either case, running the variational argument with the admissible function g or g_2 will yield the conclusion. $\square$

Remark 3.6.4 A conclusion of the proof is that

$$v(x) = \lim_{Q_i \to x, x \notin Q_i} \hat{v}(Q_i) \tag{3.30}$$

whenever x can be approximated by Q_i with $\rho(Q_i) > 0$. If $Q_i \subset \text{int}_{\overline{\Omega}}(\alpha_{t_0})$, where $t_0 \notin \mathcal{T}$, by Lemma 3.6.2 we can define $\hat{v}(Q_i)$ to be equal to the constant value of

v on $\partial_{\overline{\Omega}} V \cap S$, where V is a component of $\mathrm{int}_{\overline{\Omega}}(\alpha_{t_0})$. Thus, (3.30) can be used to define v at all points $x \in S$.

The key step in the proofs of Lemma 3.6.2 and Lemma 3.6.3 is to verify (3.25). We see that the following general statement holds (using the notation of the proof of Lemma 3.6.2):

Lemma 3.6.5 *The following statements are true.*

(a) *Let $x \in S$ and $\varepsilon > 0$. Then there exists a small $r > 0$ such that the following hold. Suppose that Q_{i_a}, Q_{i_b} are peripheral disks contained in $B(x, r)$ with $\rho(Q_{i_a}) > 0$, $\rho(Q_{i_b}) > 0$. Moreover, let $\gamma_s \subset \alpha_s$ be a path that connects Θ_2 to Q_{i_a} and $\tilde{\gamma}_t \subset \alpha_t$ be a path that connects Θ_4 to Q_{i_b}, where $s, t \in \mathcal{T}$. Then*

$$\ell_\rho(\gamma_s) + \ell_\rho(\tilde{\gamma}_t) \geq D(u) - \varepsilon.$$

In the particular case that $x \in \Theta_2$, we obtain the stronger conclusion $\ell_\rho(\tilde{\gamma}_t) \geq D(u) - \varepsilon$ and in the case $x \in \Theta_4$ we have $\ell_\rho(\gamma_s) \geq D(u) - \varepsilon$.

(b) *Let $\varepsilon > 0$, $i_0 \in \mathbb{N}$, and $x, y \in \partial Q_{i_0}$. Then there exists a small $r > 0$ such that the following hold. Suppose that $Q_{i_a} \subset B(x, r)$, $Q_{i_b} \subset B(y, r)$ with $\rho(Q_{i_a}) > 0$, $\rho(Q_{i_b}) > 0$. Moreover, let $\gamma_s \subset \alpha_s$ be a path that connects Θ_2 to Q_{i_a} and $\tilde{\gamma}_t \subset \alpha_t$ be a path that connects Θ_4 to Q_{i_b}, where $s, t \in \mathcal{T}$. Then*

$$\ell_\rho(\gamma_s) + \rho(Q_{i_0}) + \ell_\rho(\tilde{\gamma}_t) \geq D(u) - \varepsilon.$$

Proof Part (b) is proved with a slight modification of the proof of Case 0 in Lemma 3.6.2; see (3.27) and Fig. 2.1. The first claim in part (a) of the lemma follows essentially from the arguments in the proof of Lemma 3.6.3. We will only justify the latter claim in (a) for $x \in \Theta_2$, since the case $x \in \Theta_4$ is analogous. The notation that we use is the same as in the proof of Lemma 3.6.2. Suppose that $x \in \Theta_2$ and consider a test function $\zeta \in W^{1,2}(S)$ with $\zeta \equiv 1$ in a small ball $B(x, r)$ and with small Dirichlet energy $D(\zeta)$. This is provided as usual by Lemma 3.13.1. Consider the function $g_0 = (U_t \phi_t) \vee \zeta$. There exists a simple path $\gamma \subset \overline{\Omega}$ that separates Θ_1 and Θ_3 with $g_0 = 1$ on $\gamma \cap S$. We let W be the component of $\overline{\Omega} \setminus \gamma$ that contains Θ_3. If $g_0 = 1$ on points of Θ_3, then $g_0 = 0$ on Θ_1 (by choosing a ζ with a sufficiently small support), so we can set $g = 1$ on $S \cap W$ and $g = g_0$ on $S \setminus W$. If $g_0 = 1$ on points of Θ_1 then we "flip" the function g. We set $g_1 = 1$ on $S \setminus W$ and $g_1 = g_0$ on $S \cap W$. Then $g_2 := 1 - g_1$. In either case, running the standard variational argument with the admissible function g or g_2 will yield the conclusion.   □

Lemma 3.6.6 *We have $v \equiv 0$ on Θ_2 and $v \equiv D(u)$ on Θ_4. Furthermore, for each $i \in \mathbb{N}$ we have $\mathrm{osc}_{Q_i}(v) = \mathrm{osc}_{Q_i}(u) = \rho(Q_i)$. In fact,*

$$\mathrm{diam}(v(\partial Q_i \cap \alpha_t)) = \underset{\partial Q_i \cap \alpha_t}{\mathrm{osc}}(v) = \rho(Q_i)$$

for all $t \in \mathcal{T}$ with $Q_i \cap \alpha_t \neq \emptyset$.

Recall here that $\operatorname{osc}_{Q_i}(v) = \sup_{\partial Q_i}(v) - \inf_{\partial Q_i}(v)$.

Proof Let $a \in \Theta_2 \cap \alpha_t$ for some $t \in \mathcal{T}$. Then, we have $\rho(Q_i) > 0$ whenever $Q_i \cap \alpha_t \neq \emptyset$. By Lemma 3.3.7(b) we can find a sequence of peripheral disks Q_i converging to a with $Q_i \cap \alpha_t \neq \emptyset$. Using (3.30) and Lemma 3.6.1 we obtain

$$v(a) = \lim_{Q_i \to a, Q_i \cap \alpha_t \neq \emptyset} \hat{v}(Q_i) = \lim_{Q_i \to a, Q_i \cap \alpha_t \neq \emptyset} \sum_{j: Q_j \cap \alpha_t^i \neq \emptyset} \rho(Q_j). \tag{3.31}$$

This limit is equal to zero, because the sum $\sum_{j:Q_j \cap \alpha_t \neq \emptyset} \rho(Q_j)$ is convergent and the characteristic function of $\{j : Q_j \cap \alpha_t^i \neq \emptyset\}$ converges to 0 as $Q_i \to a$, $Q_i \cap \alpha_t \neq \emptyset$; recall that α_t is a path. Hence, $v(a) = 0$ in this case. If $t \notin \mathcal{T}$ and $a \in \Theta_2 \cap \alpha_t$, by Lemma 3.6.2, it suffices to show that $v(a) = 0$ whenever a can be approximated by a sequence of peripheral disks Q_i with $\rho(Q_i) > 0$. If $v(a) = 10\varepsilon > 0$ then we can find a small ball $B(a, r)$ and a peripheral disk $Q_{i_a} \subset B(a, r)$ with $\rho(Q_{i_a}) > 0$ and $\hat{v}(Q_{i_a}) > 9\varepsilon$. Let $\gamma_t, \tilde{\gamma}_t \subset \alpha_t$ for $t \in \mathcal{T}$ be paths as in the proof of Lemma 3.6.2 that connect Q_{i_a} to Θ_2, Θ_4, respectively. We have

$$\hat{v}(Q_{i_a}) + \rho(Q_{i_a}) + \ell_\rho(\tilde{\gamma}_t) = \ell_\rho(\gamma_t) + \rho(Q_{i_a}) + \ell_\rho(\tilde{\gamma}_t) = D(u),$$

which implies that $\hat{v}(Q_{i_a}) \leq D(u) - \ell_\rho(\tilde{\gamma}_t)$. However, by Lemma 3.6.5(a) we have

$$\ell_\rho(\tilde{\gamma}_t) \geq D(u) - \varepsilon$$

and this leads to a contradiction.

Now, if $a \in \Theta_4 \cap \alpha_t$ for $t \in \mathcal{T}$, again by (3.30) and Lemma 3.6.1 we have

$$v(a) = \lim_{Q_i \to a, Q_i \cap \alpha_t \neq \emptyset} \hat{v}(Q_i) = \lim_{Q_i \to a, Q_i \cap \alpha_t \neq \emptyset} \sum_{j: Q_j \cap \alpha_t^i \neq \emptyset} \rho(Q_j) = D(u),$$

where the last equality follows from Theorem 3.5.2. If $t \notin \mathcal{T}$ and $a \in \Theta_4 \cap \alpha_t$ then an application of Lemma 3.6.5(a) proves that $v(a) = D(u)$ as before.

Next, if $\rho(Q_{i_0}) = 0$ then $Q_{i_0} \subset \alpha_t$ for some $t \notin \mathcal{T}$ and Lemma 3.6.2 implies that v is constant on ∂Q_{i_0}, so $\operatorname{osc}_{Q_{i_0}}(u) = \operatorname{osc}_{Q_{i_0}}(v) = 0$. We assume that $\rho(Q_{i_0}) > 0$. If $\partial Q_{i_0} \cap \alpha_t \neq \emptyset$ for some $t \in \mathcal{T}$, then $\partial Q_{i_0} \cap \alpha_t$ contains precisely two points: x_2, which is the entry point of α_t into Q_{i_0}, and x_4, which is the exit point, as α_t travels from Θ_2 to Θ_4. Using (3.30) and Lemma 3.6.1 as in (3.31), it is easy to see that $v(x_2) = \hat{v}(Q_{i_0})$ and $v(x_4) = \hat{v}(Q_{i_0}) + \rho(Q_{i_0})$. This shows the last part of the lemma.

Now, if $a \in \partial Q_{i_0} \cap \alpha_{t_0}$ for some $t_0 \notin \mathcal{T}$ then we need to show that $\hat{v}(Q_{i_0}) \leq v(a) \leq \hat{v}(Q_{i_0}) + \rho(Q_{i_0})$. This will complete the proof of the statement that $\operatorname{osc}_{Q_{i_0}}(v) = \rho(Q_{i_0})$. Assume that $v(a) < \hat{v}(Q_{i_0}) - 10\varepsilon$ for some $\varepsilon > 0$, and without loss of generality assume that $v(a)$ is defined as in (3.21), thanks to Lemma 3.6.2. Then we can find some $Q_{i_a} \subset B(a, r)$ (where $r > 0$ is arbitrarily small) with

$\rho(Q_{i_a}) > 0$ such that $\hat{v}(Q_{i_a}) < \hat{v}(Q_{i_0}) - 9\varepsilon$. Consider the smallest open path $\gamma_s \subset \alpha_s$ that connects Θ_2 to Q_{i_a} for some $s \in \mathcal{T}$, so $\hat{v}(Q_{i_a}) = \ell_\rho(\gamma_s)$. Also, consider the smallest open path $\tilde{\gamma}_t \subset \alpha_t$ that connects Q_{i_0} to Θ_4, for some $t \in \mathcal{T}$ (this exists because $\rho(Q_{i_0}) > 0$). In particular, $\hat{v}(Q_{i_0}) + \rho(Q_{i_0}) + \ell_\rho(\tilde{\gamma}_t) = D(u)$. The path $\tilde{\gamma}_t$ lands at a point $y \in \partial Q_{i_0}$, and we can find peripheral disks Q_{i_b} intersecting $\tilde{\gamma}_t$ and lying arbitrarily close to y, by Lemma 3.3.7(b). Then by Lemma 3.6.5(b) we have

$$\ell_\rho(\gamma_s) + \rho(Q_{i_0}) + \ell_\rho(\tilde{\gamma}_t) \geq D(u) - \varepsilon$$

and this contradicts as usual the assumption $\hat{v}(Q_{i_a}) < \hat{v}(Q_{i_0}) - 9\varepsilon$.   □

Let us record a corollary:

Corollary 3.6.7 *If $t \in \mathcal{T}$, then v is strictly increasing on $S \cap \alpha_t$, in the sense that if $x, y \in S \cap \alpha_t$ and α_t hits x before hitting y as it travels from Θ_2 to Θ_4, then $v(x) < v(y)$.*

Proof Observe that $\rho(Q_i) > 0$ for all $Q_i \cap \alpha_t \neq \emptyset$, and that "between" any two points $x, y \in S \cap \alpha_t$ there exists some peripheral disk Q_i with $Q_i \cap \alpha_t \neq \emptyset$. To make this more precise, denote by $[x, y]$ the arc of α_t from x to y. Then $\mathcal{H}^1([x, y] \cap S) = 0$ by Lemma 3.5.11, so there exists a peripheral disk Q_i with $Q_i \cap [x, y] \neq \emptyset$; see also Lemma 3.3.7. Using Remark 3.6.4 and taking limits along peripheral disks that intersect α_t we obtain

$$v(x) \leq \hat{v}(Q_i) \leq v(y) - \rho(Q_i) < v(y).$$

This completes the proof.   □

The function v also satisfies a version of an upper gradient inequality. Recall that u satisfies the upper gradient inequality in Definition 3.4.1, where we exclude a family of curves Γ_0 with $\mathrm{mod}(\Gamma_0) = 0$, i.e., vanishing carpet modulus.

Lemma 3.6.8 *There exists a family of paths Γ_0 with $\mathrm{mod}_2(\Gamma_0) = 0$ such that for every path $\gamma \subset \Omega$ with $\gamma \notin \Gamma_0$ and for every open subpath β of γ we have*

$$|v(a) - v(b)| \leq \sum_{i:\overline{Q_i} \cap \beta \neq \emptyset} \rho(Q_i). \tag{3.32}$$

for all $a, b \in \overline{\beta} \cap S$.

We remark that in the sum we are using peripheral disks whose *closure* intersects β, in contrast to Definition 3.4.1. Also, here we are excluding a path family of conformal modulus zero, instead of carpet modulus. This is only a technicality and does not affect the ideas used in the proof. For the proof we will need the following technical lemma that we prove right after.

Lemma 3.6.9 *Let $\{\lambda(Q_i)\}_{i \in \mathbb{N}}$ be a sequence in $\ell^2(\mathbb{N})$. Then there exists an exceptional path family Γ in Ω with $\mathrm{mod}_2(\Gamma) = 0$ such that for all non-constant* *paths $\gamma \subset \Omega$ with endpoints in $S°$ and $\gamma \notin \Gamma$ we have*

$$\frac{1}{\delta} \sum_{i:Q_i \subset N_\delta(\gamma)} \lambda(Q_i) \operatorname{diam}(Q_i) \to 0 \quad and \quad \sum_{i \in D_\delta(\gamma)} \lambda(Q_i) \to 0$$

as $\delta \to 0$, where $D_\delta(\gamma)$ is the family of indices $i \in \mathbb{N}$ such that $\operatorname{diam}(Q_i) \geq \delta$, $Q_i \cap N_\delta(\gamma) \neq \emptyset$, and $\overline{Q}_i \cap \gamma = \emptyset$.

Proof of Lemma 3.6.8 Consider a path $\gamma \subset \Omega$ with $\mathcal{H}^1(\gamma \cap S) = 0$. This holds for mod$_2$-a.e. $\gamma \subset \Omega$ because $\mathcal{H}^2(S) = 0$. Let β be an open subpath of γ and assume that a, b are its endpoints. We wish to show that

$$|v(a) - v(b)| \leq \sum_{i:\overline{Q}_i \cap \beta \neq \emptyset} \rho(Q_i). \tag{3.33}$$

Recall that $\rho(Q_i) = \operatorname{osc}_{Q_i}(v)$ by Lemma 3.6.6. We suppose that

$$\sum_{i:\overline{Q}_i \cap \beta \neq \emptyset} \rho(Q_i) < \infty,$$

otherwise the statement is trivial. The statement is also trivial if a and b lie on the same peripheral circle ∂Q_i and β intersects $\overline{Q}_i$, so we assume that this is not the case. If $a \in \partial Q_{i_a}$ for some $i_a \in \mathbb{N}$, let $a' \in \partial Q_{i_a}$ be the last exit point of β from ∂Q_{i_a}, assuming that it is parametrized to run from a to b. Similarly, consider the point $b' \in \partial Q_{i_b}$ of first entry of β in ∂Q_{i_b}, in the case $b \in \partial Q_{i_b}$. Note that $|v(a) - v(a')| \leq \operatorname{osc}_{Q_{i_a}}(v)$ and $|v(b) - v(b')| \leq \operatorname{osc}_{Q_{i_b}}(v)$, so it suffices to prove the statement for the open subpath of β that connects a' and b'. This subpath has the property that it does not intersect the peripheral disks that possibly contain a and b on their boundary. For simplicity we denote a' by a, b' by b and the subpath by β.

By Lemma 3.3.7 we can find arbitrarily close to a peripheral disks Q_i with $Q_i \cap \beta \neq \emptyset$. Using now Lemma (b)(b), one can see that arbitrarily close to a there exist points $a'' \in \beta \cap S°$. Similarly, arbitrarily close to b there exist points $b'' \in \beta \cap S°$. By the continuity of v, it suffices to prove the statement for a'', b'' and the subpath of β that connects them instead.

Summarizing, we have reduced the proof to the case of two points $a, b \in S°$ and a subpath β of γ that connects them. We will prove this using a variational argument, very similar to the one used in Lemma 3.6.2. Since the technical details are similar and we only wish to demonstrate the new idea in this proof we assume that the points a, b can be both approximated by peripheral disks Q_i with $\rho(Q_i) > 0$, which corresponds to Case 1 in the proof of Lemma 3.6.2. (If this is not the case, and e.g. a lies in a component V of $\operatorname{int}_{\overline{\Omega}}(\alpha_{t_0})$ for some $t_0 \notin \mathcal{T}$, then one can use a "bridge" $\tau \subset V$, to connect a to a point on $\partial_{\overline{\Omega}} V \cap S$ that can be approximated by Q_i with $\rho(Q_i) > 0$. Such a bridge τ was also employed in Case 1 in the proof of Lemma 3.6.2.)

Let ℓ denote the sum in the right hand side of (3.33). If the conclusion fails, then there exists $\varepsilon > 0$ such that, say, $v(b) - v(a) \geq 10\varepsilon + \ell$. Using Lemma 3.13.1, for a small $\eta > 0$ consider a function $\zeta \in \mathcal{W}^{1,2}(S)$ that vanishes on $\partial\Omega$ with $0 \leq \zeta \leq 1$, such that $\zeta \equiv 1$ on small disjoint balls $B(a, r) \cup B(b, r) \subset \Omega$, and $D(\zeta) < \eta$. Then we can find peripheral disks $Q_{i_a} \subset B(a, r)$, $Q_{i_b} \subset B(b, r)$ with $\rho(Q_{i_a}), \rho(Q_{i_b}) > 0$ such that

$$\hat{v}(Q_{i_b}) - \hat{v}(Q_{i_a}) > 9\varepsilon + \ell. \tag{3.34}$$

Consider $s, t \in \mathcal{T}$ such that for the smallest open subpaths of γ_s, γ_t of α_s, α_t that connect Θ_2 to Q_{i_a}, Q_{i_b}, respectively, we have $\hat{v}(Q_{i_a}) = \ell_\rho(\gamma_s)$ and $\hat{v}(Q_{i_b}) = \ell_\rho(\gamma_t)$; see Lemma 3.6.1. If $\tilde{\gamma}_t$ denotes the smallest open subpath of α_t that connects Θ_4 to Q_{i_b} we have

$$\hat{v}(Q_{i_a}) - \hat{v}(Q_{i_b}) \geq \ell_\rho(\gamma_s) + \ell_\rho(\tilde{\gamma}_t) - D(u)$$

by Theorem 3.5.2. Hence, in order to obtain a contradiction to (3.34), it suffices to prove

$$\ell_\rho(\gamma_s) + \ell_\rho(\tilde{\gamma}_t) \geq D(u) - \varepsilon - \ell. \tag{3.35}$$

We will construct an admissible function g with the same procedure and notation as in Case 1 of Lemma 3.6.2. Recall the definition of U_s and U_t in (3.26). Moreover, ϕ_s is a bump function supported in a neighborhood of a strip $\Omega_{s,h} \subset A_{s-h,s+h}$ that connects Θ_2 to Q_{i_a}, and ϕ_t is a bump function supported in a neighborhood of a strip $\tilde{\Omega}_{t,h} \subset A_{t-h,t+h}$ that connects Θ_4 to Q_{i_b}. Also, for small $\delta > 0$ consider the function $\psi(x) = \max\{1 - \delta^{-1}\,\text{dist}(x, \beta), 0\}$. Now, we define

$$g_0 = (U_s\phi_s + U_t\phi_t) \vee \zeta \vee \psi$$

on S. As before, we can find a simple path on which we have $g_0 \equiv 1$ such that it separates Θ_1 from Θ_3. If W is the component that contains Θ_3, we define $g = 1$ on $S \cap W$ and $g = g_0$ on $S \setminus W$.

Recall the definitions of the index sets $F_{s,h}, F_{t,h}$ and N_h. We have

$$\underset{Q_i}{\text{osc}}(g) \leq \underset{Q_{i_t}}{\text{osc}}(\zeta) + \underset{Q_i}{\text{osc}}(\psi) + \begin{cases} 1, & i \in F_{s,h} \cup F_{t,h} \\ \rho(Q_i)/2h, & i \in N_h \\ 0, & i \notin F_{s,h} \cup F_{t,h} \cup N_h. \end{cases}$$

Since ψ is $(1/\delta)$−Lipschitz, we have

$$\underset{Q_i}{\text{osc}}(\psi) \leq \min\{\delta^{-1}\,\text{diam}(Q_i \cap N_\delta(\beta)), 1\}. \tag{3.36}$$

Testing the minimizing property of u against g (see also (3.10)) we obtain

$$D(u) \le \sum_{i \in \mathbb{N}} \rho(Q_i) \operatorname*{osc}_{Q_i}(g)$$

$$\le \sum_{i \in \mathbb{N}} \rho(Q_i) \operatorname*{osc}_{Q_i}(\zeta) + \sum_{i \in F_{s,h} \cup F_{t,h}} \rho(Q_i) + \frac{1}{2h} \sum_{i \in N_h} \rho(Q_i)^2$$

$$+ \sum_{i : Q_i \cap N_\delta(\beta) \ne \emptyset} \rho(Q_i) \operatorname*{osc}_{Q_i}(\psi).$$

Letting $h \to 0$ and choosing a small η so that $\sum_{i \in \mathbb{N}} \rho(Q_i) \operatorname{osc}_{Q_i}(\zeta) < \varepsilon$ we obtain

$$D(u) \le \varepsilon + \ell_\rho(\gamma_s) + \ell_\rho(\tilde\gamma_t) + \sum_{i : Q_i \cap N_\delta(\beta) \ne \emptyset} \rho(Q_i) \operatorname*{osc}_{Q_i}(\psi).$$

To prove our claim in (3.35), it suffices to show that the limit of the latter term as $\delta \to 0$ stays below ℓ. Note that we can split this term as

$$\sum_{i : \overline{Q}_i \cap \beta \ne \emptyset} \rho(Q_i) \operatorname*{osc}_{Q_i}(\psi) + \sum_{\substack{i : \overline{Q}_i \cap \beta = \emptyset \\ Q_i \cap N_\delta(\beta) \ne \emptyset}} \rho(Q_i) \operatorname*{osc}_{Q_i}(\psi).$$

Using (3.36) we see that the first term is already bounded by ℓ, so it suffices to show that the second term converges to 0 as $\delta \to 0$. Consider the family $D_\delta(\beta)$ of indices $i \in \mathbb{N}$ such that $\operatorname{diam}(Q_i) \ge \delta$, $Q_i \cap N_\delta(\beta) \ne \emptyset$, and $\overline{Q}_i \cap \beta = \emptyset$, as in the statement of Lemma 3.6.9. Observe that if $Q_i \cap N_\delta(\beta) \ne \emptyset$ and $\operatorname{diam}(Q_i) < \delta$, then $Q_i \subset N_{2\delta}(\beta)$. Hence, using (3.36) it suffices to show that

$$\frac{1}{\delta} \sum_{i : Q_i \subset N_{2\delta}(\beta)} \rho(Q_i) \operatorname{diam}(Q_i) \to 0 \quad \text{and} \quad \sum_{i \in D_\delta(\beta)} \rho(Q_i) \to 0$$

as $\delta \to 0$.

This follows immediately from Lemma 3.6.9. Indeed, the family of paths γ for which the conclusion of the lemma fails has conformal modulus equal to zero. Since the family of paths γ that have a subpath β for which the conclusion of the lemma fails also has conformal modulus zero, this completes the proof of the upper gradient inequality. $\square$

Proof of Lemma 3.6.9 By the subadditivity of modulus, we can treat each of the claims separately. Fix a curve γ with $\mathcal{H}^1(\gamma \cap S) = 0$. The latter holds for mod_2-a.e. $\gamma \subset \Omega$.

We can cover $N_\delta(\gamma)$ by balls $B_{j,\delta}$ of radius 2δ centered at γ such that $\frac{1}{20} B_{j,\delta}$ are disjoint. To do this, one can cover γ by balls of radius $\delta/10$ and extract a disjoint subcollection $\{B_l\}_l$ such that the balls $5B_l$ still cover γ; see for instance

[21, Theorem 1.2]. We now define $\{B_{j,\delta}\}_j$ to be the collection of balls $\{20B_l\}_l$, each of which has radius 2δ. Then any point $x \in N_\delta(\gamma)$ is δ-far from γ and thus $\delta + \delta/2$-far from the center of a ball $5B_l$ (of radius $\delta/2$), which is equal to $\frac{1}{4}B_{j,\delta}$ for some j. It follows that $x \in B_{j,\delta}$.

Next, consider the subfamily $\{B'_{j,\delta}\}_j$ of the balls $\{B_{j,\delta}\}_j$ that are not entirely contained in any peripheral disk. We note that $\{B'_{j,\delta}\}_j$ covers $\bigcup_{i:Q_i \subset N_\delta(\gamma)} Q_i$ and $\gamma \cap S$. In fact, as $\delta \to 0$ along a sequence one can construct covers $\{B'_{j,\delta}\}_j$ as above such that $\bigcup_j B'_{j,\delta}$ is decreasing. Furthermore, as $\delta \to 0$ we have that

$$\gamma \cap \left(\bigcup_j B'_{j,\delta}\right) \to \gamma \cap S. \tag{3.37}$$

Indeed, if this failed, then there would exist $i_0 \in \mathbb{N}$ and some $x \in Q_{i_0} \cap \gamma$ which belongs to $\bigcup_j B'_{j,\delta}$ infinitely often as $\delta \to 0$. This contradicts the construction of the cover $\{B'_{j,\delta}\}$, since $\text{dist}(x, \partial Q_{i_0}) > 0$, and a ball $B'_{j,\delta}$ that contains x would be entirely contained in Q_{i_0} for small δ, so it would have been discarded during the construction.

Now consider the function $\Lambda(x) = \sum_{i \in \mathbb{N}} \frac{\lambda(Q_i)}{\text{diam}(Q_i)} \chi_{Q_i}(x)$. Observe that by the quasiroundness assumption (3.1) there exists a constant $C > 0$ such that

$$\frac{1}{C} \int_{Q_i} \Lambda(x) \, d\mathcal{H}^2(x) \le \lambda(Q_i) \, \text{diam}(Q_i) \le C \int_{Q_i} \Lambda(x) \, d\mathcal{H}^2(x), \tag{3.38}$$

and

$$\frac{1}{C} \int_{Q_i} \Lambda(x)^2 \, d\mathcal{H}^2(x) \le \lambda(Q_i)^2 \le C \int_{Q_i} \Lambda(x)^2 \, d\mathcal{H}^2(x), \tag{3.39}$$

for all $i \in \mathbb{N}$. Using the properties of the cover $\{B'_{j,\delta}\}_j$ and the uncentered maximal function $M\Lambda$ (see [21, Sect. 2]) we have:

$$\sum_{i:Q_i \subset N_\delta(\gamma)} \lambda(Q_i) \, \text{diam}(Q_i) \le C \int_{\bigcup_{i:Q_i \subset N_\delta(\gamma)} Q_i} \Lambda(x) \, d\mathcal{H}^2(x)$$

$$\le C \int_{\bigcup_j B'_{j,\delta}} \Lambda(x) \, d\mathcal{H}^2(x)$$

$$\le C \sum_j \int_{B'_{j,\delta}} \Lambda(x) \, d\mathcal{H}^2(x)$$

$$\le C'\delta \sum_j \inf_{x \in \frac{1}{20}B'_{j,\delta}} M\Lambda(x)$$

$$\leq C'' \delta \sum_j \int_{\gamma \cap (\frac{1}{20} B'_{j,\delta})} M\Lambda(x) \, d\mathcal{H}^1(x)$$

$$= C'' \delta \int_{\gamma \cap (\bigcup_j \frac{1}{20} B'_{j,\delta})} M\Lambda(x) \, d\mathcal{H}^1(x)$$

$$\leq C'' \delta \int_{\gamma \cap (\bigcup_j B'_{j,\delta})} M\Lambda(x) \, d\mathcal{H}^1(x).$$

The first part of the lemma will follow, if we show that

$$\int_{\gamma \cap (\bigcup_j B'_{j,\delta})} M\Lambda(x) \, d\mathcal{H}^1(x) \to 0 \tag{3.40}$$

as $\delta \to 0$ for mod_2-a.e. curve γ. First observe that $M\Lambda \in L^2(\Omega)$ by the maximal theorem (see [21, Theorem 2.2]), since $\Lambda \in L^2(\Omega)$ by (3.39). Hence, $\int_\gamma M\Lambda(x) \, d\mathcal{H}^1(x) < \infty$ for mod_2-a.e. $\gamma \subset \Omega$. By construction, $\gamma \cap (\bigcup_j B'_{j,\delta})$ decreases to $\gamma \cap S$ with $\mathcal{H}^1(\gamma \cap S) = 0$. The dominated convergence theorem now immediately implies (3.40).

Now, we show the second part of the lemma. Recall that $D_\delta(\gamma)$ contains all indices $i \in \mathbb{N}$ for which $\mathrm{diam}(Q_i) \geq \delta$, $Q_i \cap N_\delta(\gamma) \neq \emptyset$, and $\overline{Q}_i \cap \gamma = \emptyset$. We wish to show that the family of paths for which the conclusion fails has conformal modulus equal to zero. We first remark that this family contains no constant paths, by assumption. By the subadditivity of conformal modulus it suffices to show that for every $d > 0$, $\varepsilon_0 > 0$ the family Γ of paths γ, having endpoints in S°, with $\mathrm{diam}(\gamma) \geq d$ and

$$\limsup_{\delta \to 0} \sum_{i \in D_\delta(\gamma)} \lambda(Q_i) \geq \varepsilon_0,$$

has conformal modulus zero. Let $\{\lambda_0(Q_i)\}_{i \in \mathbb{N}}$ be a finitely supported sequence with $\lambda_0(Q_i) = \lambda(Q_i)$ or $\lambda_0(Q_i) = 0$, for each $i \in \mathbb{N}$. Then

$$\sum_{i \in D_\delta(\gamma)} \lambda_0(Q_i) \to 0$$

as $\delta \to 0$, since $\overline{Q}_i \cap \gamma = \emptyset$ and thus $\mathrm{dist}(\overline{Q}_i, \gamma) > 0$ for all $i \in D_\delta(\gamma)$. Here, it is crucial that γ has endpoints in S°, and the preceding statement would fail if γ was an open path and one of its endpoints was on a peripheral circle. Consequently, if $\gamma \in \Gamma$, then

$$\limsup_{\delta \to 0} \sum_{i \in D_\delta(\gamma)} h(Q_i) \geq \varepsilon_0, \tag{3.41}$$

where $h(Q_i) := \lambda(Q_i) - \lambda_0(Q_i)$.

We will construct an admissible function $\tilde{h}$ for $\text{mod}_2(\Gamma)$ with arbitrarily small mass. By the summability assumption on λ, for each $\eta > 0$ we can find λ_0 as above such that $\sum_{i \in \mathbb{N}} h(Q_i)^2 < \eta$. For each $i \in \mathbb{N}$ consider balls $B(x_i, r_i) \subset Q_i \subset B(x_i, R_i)$ as in the quasiroundness assumption (3.1) with $R_i = \text{diam}(Q_i)$. We define

$$\tilde{h} = c_0 \sum_{i \in \mathbb{N}} \frac{h(Q_i)}{R_i} \chi_{B(x_i, 4R_i)}$$

where c_0 is a constant to be determined, independent of η. Note that by Lemma 3.3.2 and the fact that the balls $B(x_i, R_i/K_0) \subset B(x_i, r_i)$ are disjoint we have

$$\int \tilde{h}(x)^2 \, d\mathcal{H}^2(x) \le C c_0^2 \sum_{i \in \mathbb{N}} \frac{h(Q_i)^2}{R_i^2} R_i^2 \le C c_0^2 \eta.$$

Since this can be made arbitrarily small, it remains to show that $\tilde{h}$ is admissible for Γ.

Fix a curve $\gamma \in \Gamma$, so (3.41) holds. Observe that $\max_{i \in D_\delta(\gamma)} \text{diam}(Q_i) \to 0$ as $\delta \to 0$, by the definition of $D_\delta(\gamma)$ and the fact that there are only finitely many peripheral disks with "large" diameter. Now, let δ be sufficiently small, so that $8 \max_{i \in D_\delta(\gamma)} \text{diam}(Q_i) < d \le \text{diam}(\gamma)$ and $\sum_{i \in D_\delta(\gamma)} h(Q_i) > \varepsilon_0/2$. If $i \in D_\delta(\gamma)$ then $R_i = \text{diam}(Q_i) \ge \delta$ and $Q_i \cap N_\delta(\gamma) \ne \emptyset$, thus $B(x_i, 2R_i)$ meets γ. By the choice of δ, γ has to exit $B(x_i, 4R_i)$. Hence, $\mathcal{H}^1(\gamma \cap B(x_i, 4R_i)) \ge 2R_i$, which implies that

$$\int_\gamma \tilde{h}(x) \, d\mathcal{H}^1(x) \ge c_0 \sum_{i \in D_\delta(\gamma)} \frac{h(Q_i)}{R_i} \mathcal{H}^1(\gamma \cap B(x_i, 4R_i)) \ge c_0 \varepsilon_0.$$

We choose $c_0 = 1/\varepsilon_0$ and this completes the proof. $\qquad\qquad\square$

Remark 3.6.10 It is clear from the proof that this general lemma holds for carpets S of area zero for which the peripheral disks are uniformly quasiround; the Ahlfors regularity assumption was not used here.

Remark 3.6.11 If we could prove that the function v satisfies the same upper gradient inequality as u (see Definition 3.4.1) then $x \mapsto v(x)/D(u)$ would be admissible for the free boundary problem with respect to Θ_2 and Θ_4. One then could show that $v/D(u)$ is carpet-harmonic and thus v is carpet-harmonic. We believe that the validity of other types of the upper gradient inequality for v (among the possibilities of excluding paths of 2-modulus zero or carpet modulus zero and using paths that intersect Q_i or $\overline{Q}_i$) depends on the geometry of the peripheral disks and their separation. Without any extra geometric assumptions it is unclear how to establish the harmonicity of v.

3.7 Definition of f

Let $D = D(u) = \sum_{i \in \mathbb{N}} \rho(Q_i)^2$, and consider the continuous function

$$f = (u, v) \colon S \to [0, 1] \times [0, D].$$

The fact that the range of f is $[0, 1] \times [0, D]$ is justified by Lemma 3.6.6. The same lemma also implies that $f(\partial\Omega) = \partial([0, 1] \times [0, D]) = \partial S_0$, where $S_0 := \mathbb{C} \setminus [0, 1] \times [0, D]$. If $\rho(Q_i) = 0$, then u and v are constant on ∂Q_i, so $f(\partial Q_i)$ is a single point, denoted by $\overline{S}_i$ or ∂S_i. If $\rho(Q_i) > 0$, again by Lemma 3.6.6 we have

$$f(\partial Q_i) \subset [m_{Q_i}(u), M_{Q_i}(u)] \times [m_{Q_i}(v), M_{Q_i}(v)] =: \overline{S}_i, \tag{3.42}$$

where $M_{Q_i}(u) - m_{Q_i}(u) = M_{Q_i}(v) - m_{Q_i}(v) = \rho(Q_i)$. Thus, the image of ∂Q_i is contained in a square of side length $\rho(Q_i)$. We define S_i to be the open square $(m_{Q_i}(u), M_{Q_i}(u)) \times (m_{Q_i}(v), M_{Q_i}(v)) = \operatorname{int}(\overline{S}_i)$, or the empty set in the case $\rho(Q_i) = 0$. If $\rho(Q_i) > 0$, then we will call $\overline{S}_i$ a *non-degenerate* square. We claim that these squares have disjoint interiors:

Lemma 3.7.1 *The (open) squares S_i, $i \in \mathbb{N}$, are disjoint. Furthermore, for each $i \in \mathbb{N}$ we have $S_i \cap f(S) = \emptyset$ and $f(\partial Q_i) \subset \partial S_i$.*

Proof Assume that $\rho(Q_i), \rho(Q_j) > 0$ for some $i, j \in \mathbb{N}$, $i \neq j$, and that $S_i \cap S_j \neq \emptyset$. Since the x-coordinates of the squares intersect at an interval of positive length, there exists some $t \in \mathcal{T}$ such that $\overline{Q}_i \cap \alpha_t \neq \emptyset$ and $\overline{Q}_j \cap \alpha_t \neq \emptyset$. Assume that the path α_t meets ordered points $x_1, x_2 \in \partial Q_i$ and then $y_1, y_2 \in \partial Q_j$ as it travels from Θ_2 to Θ_4. This can be justified using the properties of α_t from Lemma 3.5.14. By Lemma 3.6.6 and Corollary 3.6.7 we have

$$v(x_1) + \rho(Q_i) = v(x_2) < v(y_1) = v(y_2) - \rho(Q_j).$$

This clearly contradicts the assumption that $S_i \cap S_j \neq \emptyset$.

For our second claim, assume that there exists some $x \in S$ with $f(x) \in S_{i_0}$ for some $i_0 \in \mathbb{N}$. If $x \in \alpha_t$ for some $t \in \mathcal{T}$, then x can be approximated by peripheral disks Q_i with $\rho(Q_i) > 0$; see Lemma 3.3.7 and recall that all peripheral disks intersecting α_t satisfy $\rho(Q_i) > 0$. By the continuity of f the diameter $\operatorname{diam}(S_i)$ becomes arbitrarily small as $Q_i \to x$, so $S_i \subset S_{i_0}$ if Q_i is sufficiently close to x. This contradicts the first part of the lemma.

If $x \in \alpha_t$ for some $t \notin \mathcal{T}$ then there exists a point $y \in \alpha_t$ such that $v(x) = v(y)$ and y can be approximated by Q_i with $\rho(Q_i) > 0$; see Lemma 3.6.2 and the discussion that precedes it. Thus $f(x) = f(y) = (t, v(y))$, and the previous case applies to yield a contradiction.

The final claim follows from (3.42) and the previous parts of the lemma. $\qquad\square$

In fact, we have the following:

Corollary 3.7.2 *For each $i \in \mathbb{N} \cup \{0\}$ we have $f(\partial Q_i) = \partial S_i$. Moreover,*

$$f(S) = [0, 1] \times [0, D] \setminus \bigcup_{i \in \mathbb{N}} S_i =: \mathcal{R}, \quad \mathcal{H}^2(\mathcal{R}) = 0,$$

and the intersection of a non-degenerate square $\overline{S}_i$, $i \in \mathbb{N}$, with $\partial S_0 = \partial([0, 1] \times [0, D])$ or with another non-degenerate square $\overline{S}_j$, $j \in \mathbb{N}$, $j \neq i$, is either the empty set or a singleton.

Proof By the preceding lemma we know that $f(\partial Q_i) \subset \partial S_i$ for each $i \in \mathbb{N}$. Consider a continuous extension $\widetilde{f} \colon \overline{\Omega} \to [0, 1] \times [0, D]$ such that $\widetilde{f}(Q_i) \subset S_i$, whenever S_i is a non-degenerate square. One way to find such an extension is to consider a Poisson extension $\tilde{v}$ of v in each peripheral disk as in Lemma 3.4.8. Then $\tilde{v}(Q_i) \subset (m_{Q_i}(v), M_{Q_i}(v))$, and also $\tilde{u}(Q_i) \subset (m_{Q_i}(u), M_{Q_i}(u))$, by the maximum principle. Hence, if we define $\widetilde{f} = (\tilde{u}, \tilde{v})$ we have the desired property $\widetilde{f}(Q_i) \subset S_i$ whenever S_i is non-degenerate. Combining this with Lemma 3.7.1, we see that for each non-degenerate S_i we have

$$\widetilde{f}(\overline{\Omega} \setminus Q_i) \subset [0, 1] \times [0, D] \setminus S_i. \tag{3.43}$$

First we show that $f(\partial \Omega) = f(\partial Q_0) = \partial S_0$. Recall that $u \equiv 0$ on Θ_1 and $u \equiv 1$ on Θ_3, so these sets are mapped into the left and right vertical sides of the rectangle ∂S_0, respectively; see Theorem 3.4.5. Also, by Lemma 3.6.6, $v \equiv 0$ on Θ_2 and $v \equiv D$ on Θ_4, so these sets are mapped to the bottom and top sides of ∂S_0, respectively. By continuity, we must have $f(\partial Q_0) = \partial S_0$.

Proposition 3.5.1 shows that the functions $f|_{\Theta_2}$, $f|_{\Theta_4}$ are increasing from 0 to 1 if Θ_2, Θ_4 are parametrized as arcs from Θ_1 to Θ_3. Thus, $\widetilde{f}|_{\partial \Omega}$ winds once around every point of $(0, 1) \times (0, D)$. By homotopy, using (3.43), we see that $\widetilde{f}|_{\partial Q_i}$ winds once around every point of $\mathrm{int}(S_i)$. Thus $\widetilde{f}(\partial Q_i) = \partial S_i$, and $\widetilde{f}(Q_i) = S_i$; see [32, Chap. II].

Note that the area of $[0, 1] \times [0, D]$ is equal to $D = \sum_{i \in \mathbb{N}} \rho(Q_i)^2$, which is the sum of the areas of the squares S_i. By Lemma 3.7.1, the squares have disjoint interiors, so in some sense they tile $[0, 1] \times [0, D]$, and this already shows that $\mathcal{H}^2(\mathcal{R}) = 0$. Furthermore, we obtain that the boundaries $\partial S_i = f(\partial Q_i)$ are dense in $[0, 1] \times [0, D] \setminus \bigcup_{i \in \mathbb{N}} S_i$. Since the sets ∂Q_i, $i \in \mathbb{N}$, are also dense in the carpet S and f is continuous, we obtain $f(S) = \mathcal{R}$.

For the last claim, assume that two squares $\overline{S}_i$, $\overline{S}_j$, $i \neq j$, share part of a vertical side, i.e., there exists a non-degenerate vertical line segment $\tau := \{t\} \times [s_1, s_2] \subset \overline{S}_i \cap \overline{S}_j$. Let $s \in (s_1, s_2)$ and consider, by surjectivity, a point $x \in \partial Q_i$ such that $f(x) = (t, s) \in \tau$. We claim that x cannot be approximated by Q_k, $k \neq i$, with $\rho(Q_k) > 0$. Indeed, if this was the case, then $f(\partial Q_k) = \partial S_k$ would be non-degenerate distinct squares that approximate the point $f(x)$ by continuity. Obviously, this cannot happen since $f(x) \in \tau$, and S_i, S_k have to be disjoint by

Lemma 3.7.1. Hence, $\rho(Q_k) = 0$ for all Q_k contained in a neighborhood of x. The upper gradient inequality of u along with continuity imply that u is constant in a neighborhood of x; see Lemma 2.7.2 for a proof. By Lemma 3.6.2 and the definition v we conclude that v is also constant in a neighborhood of x; see also the comments before Lemma 3.6.2. In particular, f is constant in some arc of ∂Q_i containing x. However, there are countably many such subarcs of ∂Q_i, and uncountably many preimages $f^{-1}((t, s))$, $s \in (s_1, s_2)$. This is a contradiction. The same argument applies if $\overline{S}_i$, $\overline{S}_j$ share a horizontal segment or if $\overline{S}_i \cap \partial S_0 \neq \emptyset$. □

Combining the upper gradient inequalities of u and v we obtain:

Proposition 3.7.3 *There exists a family of paths Γ_0 with $\mathrm{mod}_2(\Gamma_0) = 0$ such that for every path $\gamma \subset \Omega$ with $\gamma \notin \Gamma_0$ and for every open subpath β of γ we have*

$$|f(x) - f(y)| \leq \sqrt{2} \sum_{i:\overline{Q}_i \cap \beta \neq \emptyset} \rho(Q_i) = \sum_{i:\overline{Q}_i \cap \beta \neq \emptyset} \mathrm{diam}(\overline{S}_i)$$

for all $x, y \in \overline{\beta} \cap S$.

Observe that the paths β are contained in Ω, so $\overline{Q}_0 \cap \beta = \emptyset$, and we never include the term $\mathrm{diam}(\overline{S}_0) = \mathrm{diam}(\mathbb{C} \setminus (0, 1) \times (0, D)) = \infty$ in the above summations.

Proof This type of upper gradient inequality (without the $\sqrt{2}$ term) holds for the function v by Lemma 3.6.8. If show that u also satisfies this type of upper gradient inequality (without $\sqrt{2}$), then it will follow that f satisfies the desired inequality. Recall that $u \in W^{1,2}(S)$ so it satisfies the upper gradient inequality in Definition 3.4.1 with an exceptional family Γ_0 that has carpet modulus equal to 0. Lemma 3.3.1 now implies that $\mathrm{mod}_2(\Gamma_0) = 0$. To complete the proof, note that the sum over $\{i : \overline{Q}_i \cap \beta \neq \emptyset\}$ is larger than the sum over $\{i : Q_i \cap \beta \neq \emptyset\}$ (which was used in Definition 3.4.1). □

The upper gradient inequality has the next important corollary.

Corollary 3.7.4 *There exists a family of paths Γ_0 with $\mathrm{mod}_2(\Gamma_0) = 0$ such that the following holds:*
For every open path $\gamma \subset \Omega$, $\gamma \notin \Gamma_0$, with endpoints $x, y \in S$ and every Lipschitz map $\pi : \mathbb{R}^2 \to \mathbb{R}$ we have

$$|\pi(f(x)) - \pi(f(y))| \leq \mathcal{H}^1\left(\bigcup_{i:\overline{Q}_i \cap \gamma \neq \emptyset} \pi(S_i) \right) \leq \sum_{i:\overline{Q}_i \cap \gamma \neq \emptyset} \mathcal{H}^1(\pi(S_i)).$$

Remark 3.7.5 The corollary does *not* imply that for almost every γ we have $\mathcal{H}^1(f(\gamma \cap S)) = 0$; cf. Lemma 3.9.2. To illustrate its meaning, let γ be a non-exceptional path joining x and y, and π be the orthogonal projection to the line passing through $f(x)$, $f(y)$. Then the corollary says that the projection of the squares intersected by $f(\gamma)$ "covers" the entire distance from $f(x)$ to $f(y)$.

Remark 3.7.6 If $\{\lambda(Q_i)\}_{i\in\mathbb{N}}$ is a sequence in $\ell^2(\mathbb{N})$, it can be proved that the family of paths Γ in $\mathbb{C}$ satisfying

$$\sum_{i:\overline{Q}_i\cap\gamma\neq\emptyset} \lambda(Q_i) = \infty$$

has 2-modulus zero. If we were using Q_i instead of $\overline{Q}_i$ under the summation, then this would follow immediately from Lemma 3.3.1. The proof in our case is in fact exactly the same as the proof of Lemma 3.3.1; see the proof of Lemma 2.3.3.

Proof Let $\gamma \subset \Omega$ be a path such that the upper gradient inequality of Proposition 3.7.3 holds for all open subpaths β of γ and such that $\sum_{i:\overline{Q}_i\cap\gamma\neq\emptyset} \rho(Q_i) < \infty$. The latter holds for all curves outside a family conformal modulus zero, by the preceding remark.

If $x \in \partial Q_{i_x}$ for some $i_x \in \mathbb{N}$, then we let $x' \in \partial Q_{i_x}$ be the point of last exit of γ from Q_{i_x}, and we similarly consider the point $y' \in \partial Q_{i_y}$ of first entry of γ into Q_{i_y} (after x'), in the case $y \in \partial Q_{i_y}$. Since the differences $|\pi(f(x)) - \pi(f(x'))|$ and $|\pi(f(y)) - \pi(f(y'))|$ are controlled by $\mathcal{H}^1(\pi(S_{i_x}))$ and $\mathcal{H}^1(\pi(S_{i_y}))$, respectively, it suffices to prove the statement with x and y replaced by x' and y', respectively, and with γ replaced by its open subpath connecting x' and y'.

Hence, from now on we assume that γ is an open path with endpoints x, y and we suppose that γ does not intersect any closed peripheral disk $\overline{Q}_i$ with $x \in \partial Q_i$ or $y \in \partial Q_i$. Fix $\varepsilon > 0$ and consider a finite index set $J \subset \mathbb{N}$ such that

$$\sum_{\substack{i:\overline{Q}_i\cap\gamma\neq\emptyset \\ i\notin J}} \rho(Q_i) < \varepsilon. \tag{3.44}$$

Assume γ is parametrized as it runs from x to y. Using Lemma 3.3.8, we may obtain finitely many open subpaths $\gamma_1, \ldots, \gamma_m$ of γ with the following properties:

1. the upper gradient inequality of f holds along each path γ_k, $k = 1, \ldots, m$,
2. the paths γ_k intersect disjoint sets of peripheral disks $\overline{Q}_i$, $i \in \mathbb{N} \setminus J$, and
3. the path γ_1 starts at $x_1 = x$, the path γ_m terminates at $y_m = y$, and in general the path γ_k has endpoints $x_k, y_k \in S$ such that for $1 \le k \le m-1$ we either have

   - $y_k = x_{k+1}$, or
   - $y_k, x_{k+1} \in \partial Q_{j_k}$ for some $j_k \in \mathbb{N}$. The peripheral disks $\overline{Q}_{j_k}$ are distinct and they are intersected by γ.

   We denote by $I \subset \{1, \ldots, m\}$ the set of indices k for which the second alternative holds.

The assumption that γ does not intersect any closed peripheral disk $\overline{Q}_i$ with $x \in \partial Q_i$ or $y \in \partial Q_i$ is essential, otherwise property (2) could fail.

For each S_{j_k}, $k \in I$, the image $\pi(S_{j_k}) \subset \mathbb{R}$ is an interval. Thus, $\bigcup_{k\in I} \pi(S_{j_k})$ is a union of finitely many intervals that contain the points $\pi(f(y_k))$ and $\pi(f(x_{k+1}))$,

$k \in I$, in their closure. Without loss of generality, assume that $\pi(f(x)) \le \pi(f(y))$. The interval $[\pi(f(x)), \pi(f(y))]$ is covered by the union of the set $\bigcup_{k \in I} \pi(S_{j_k})$ together with the closed intervals between $\pi(f(x_k))$ and $\pi(f(y_k))$, $k = 1, \ldots, m$. If L is the Lipschitz constant of π, then we have

$$|\pi(f(x)) - \pi(f(y))| \le \mathcal{H}^1\left(\bigcup_{k \in I} \pi(S_{j_k})\right) + \sum_{k=1}^{m} |\pi(f(x_k)) - \pi(f(y_k))|$$

$$\le \mathcal{H}^1\left(\bigcup_{i:\overline{Q}_i \cap \gamma \neq \emptyset} \pi(S_i)\right) + L \sum_{k=1}^{m} |f(x_k) - f(y_k)|$$

$$\le \mathcal{H}^1\left(\bigcup_{i:\overline{Q}_i \cap \gamma \neq \emptyset} \pi(S_i)\right) + L \sum_{k=1}^{m} \sum_{i:\overline{Q}_i \cap \gamma_k \neq \emptyset} \sqrt{2}\rho(Q_i)$$

Since the paths γ_k intersect disjoint sets of peripheral disks $\overline{Q}_i$, $i \in \mathbb{N} \setminus J$, the latter term is bounded by

$$\sqrt{2}L \sum_{\substack{i:\overline{Q}_i \cap \gamma \neq \emptyset \\ i \notin J}} \rho(Q_i) < \sqrt{2}L\varepsilon,$$

where we used (3.44). Letting $\varepsilon \to 0$ finishes the proof. $\qquad\square$

3.8 Injectivity of f

We will prove that f is injective in two steps. First we show that $f: S \to \mathcal{R}$ is a *light map*, i.e., the preimage of every point $z \in \mathcal{R}$ contains no non-trivial continua. Then we show that the preimage of every point $z \in \mathcal{R}$ is actually a single point. These are also the steps that are followed in [33, Sect. 9].

From now on, f will denote the continuous extension $\tilde{f}: \overline{\Omega} \to [0, 1] \times [0, D]$, as in the proof of Corollary 3.7.2, which has the property that $\tilde{f}(Q_i) = S_i$ whenever S_i, $i \in \mathbb{N}$, is a non-degenerate square. Also, recall that the coordinates $\tilde{u}$ and $\tilde{v}$ of f are harmonic in the classical sense inside each Q_i, $i \in \mathbb{N}$, and that $\alpha_t = \tilde{u}^{-1}(t)$, $t \in [0, 1]$.

Lemma 3.8.1 *For every $z \in \mathcal{R}$ the set $f^{-1}(z)$ contains no non-trivial continua.*

The proof will follow from a modulus-type argument. Essentially, the carpet modulus of a family of curves passing through the point z is zero; however, this would not be the case for the curves passing through $f^{-1}(z)$ if the latter contains a continuum. We will use the upper gradient inequality for f in Proposition 3.7.3 and its corollary to compare the modulus in the image and the preimage. In some sense, the map f preserves carpet modulus, and this prevents a curve family of positive modulus from being mapped to a curve family of modulus zero.

Proof Assume that there exists a non-trivial continuum $E \subset f^{-1}(z)$ for some $z \in \mathcal{R}$. Note that the preimage $f^{-1}(z)$ cannot intersect both Θ_1 and Θ_3. We let $F = \Theta_l$, where $l = 1$ or $l = 3$ and Θ_l is such that $E \cap \Theta_l = \emptyset$. We consider a small $R > 0$ such that $f^{-1}(B(z, R)) \cap F = \emptyset$. Such an R exists because of the following general fact: If $g : X \to Y$ is a continuous map between metric spaces X, Y and X is compact, then for each $z \in Y$ and $\varepsilon > 0$ there exists $R > 0$ such that $g^{-1}(B(z, R)) \subset N_\varepsilon(g^{-1}(z))$.

By the structure of $\mathcal{R}$ (see Corollary 3.7.2), the point z, among other possibilities, might lie on a square ∂S_{i_1}, $i_1 \in \mathbb{N}$, or it could be the common vertex of two intersecting squares $\partial S_{i_1}, \partial S_{i_2}, i_1, i_2 \in \mathbb{N}$. We construct closed annuli A_j centered at z in the following way. Let $A_1 = \overline{A}(z; r_1, R_1) = \overline{B}(z, R_1) \setminus B(z, r_1)$, where $R_1 = R$ and $r_1 = R_1/2$. Let $R_2 < r_1$ be so small that no square S_i (except possibly for S_{i_1}, S_{i_2}) intersects both A_1 and $A_2 := \overline{A}(z; r_2, R_2)$, where $r_2 = R_2/2$. This can be achieved because the squares near z are arbitrarily small, with the possible exception of S_{i_1} and S_{i_2}. We proceed inductively to obtain annuli $A_1, \dots, A_N$, for some fixed large N; cf. proof of Lemma 3.13.1. We then replace each A_j with $A_j \cap ([0, 1] \times [0, D])$.

For $Q_i \cap f^{-1}(A_j) \neq \emptyset$ (equiv. $S_i \cap A_j \neq \emptyset$) we set $\lambda(Q_i) = \frac{1}{Nr_j} d_j(S_i)$, where $d_j(S_i) = \mathcal{H}^1(\{r \in [r_j, R_j] : S_i \cap B(z, r) \neq \emptyset\})$. In other words, $d_j(S_i)$ is the *radial diameter* of the intersection $S_i \cap A_j$. If $S_i \cap A_j = \emptyset$ we set $\lambda(Q_i) = 0$. Note that there exists a constant $C > 0$ such that

$$d_j(S_i)^2 \leq C\mathcal{H}^2(S_i \cap A_j) \tag{3.45}$$

for all $i \in \mathbb{N}$ and $j = 1, \dots, N$. This is true because, for example, the squares are uniformly Ahlfors 2-regular sets; see also Remark 3.5.10.

We now wish to construct a Lipschitz family of non-exceptional open paths in Ω that connects E to F, but avoids the peripheral disk $\overline{Q}_{i_1}$ in the case $z \in \partial S_{i_1}$, or avoids the peripheral disks $\overline{Q}_{i_1}$ and $\overline{Q}_{i_2}$ in the case $z \in \partial S_{i_1} \cap \partial S_{i_2}$. Here non-exceptional means that the paths, as well as their subpaths, do not lie in some given path family of 2-modulus zero. More specifically, we require that for all open subpaths of these non-exceptional paths the conclusion of Corollary 3.7.4 holds.

If none of the aforementioned two scenarios occur (i.e., $z \notin \partial S_{i_1}$ or $z \notin \partial S_{i_1} \cap \partial S_{i_2}$), then we consider an open path τ joining E to F. We can make now direct use of Lemma 3.3.4 to obtain a small $\delta > 0$ and non-exceptional paths β_s, for a.e. $s \in (0, \delta)$, that connect E to F. If $z \in \partial S_{i_1}$ (i.e., we are in the first of the two scenarios), then we split into two cases. In the case $E \setminus \partial Q_{i_1} \neq \emptyset$, we consider a point $x \in E \setminus \partial Q_{i_1}$ and connect it to F with a path $\tau \subset \Omega \setminus \overline{Q}_{i_1}$. Then for a small $\delta > 0$ the perturbations β_s given by Lemma 3.3.4 do not intersect $\overline{Q}_{i_1}$ and have the desired properties. In the case $E \subset \partial Q_{i_1}$, E has to contain an arc; we choose x to be an interior point (i.e., not an endpoint) of this arc. We connect $E \ni x$ to

F with an open path $\tau \subset \Omega \setminus \overline{Q}_{i_1}$; the latter region is just a topological annulus. Lemma 3.3.4 yields non-exceptional paths β_s, for a.e. $s \in (0, \delta)$, but this time the paths are not necessarily disjoint from $\overline{Q}_{i_1}$. To amend this, we consider a possibly smaller $\delta > 0$ and open subpaths of β_s that we still denote by β_s, which connect E to F without entering $\overline{Q}_{i_1}$. Finally, one has to treat the case $z \in \partial S_{i_1} \cap \partial S_{i_2}$, but this is done exactly as the case $z \in \partial S_{i_1}$.

For the moment, we fix a path β_s and we consider subpaths of β_s as follows. Assume β_s is parametrized as it runs from F to E. Let γ_j be the open subpath from the point of last entry of β_s into $f^{-1}(A_j)$ until the point of first entry into $f^{-1}(B(z, r_j))$, $j = 1, \ldots, N$. Then γ_j intersects only peripheral disks meeting $f^{-1}(A_j)$. Hence, by construction of the annuli A_j, the paths γ_j for distinct indices j intersect disjoint sets of peripheral disks.

Let $\pi : \mathbb{R}^2 \to \mathbb{R}$ be the projection $w = z + re^{i\theta} \mapsto r$, so $d_j(S_i) = \mathcal{H}^1(\pi(S_i))$. If the endpoints of γ_j lie in S, then by Corollary 3.7.4 we have:

$$\sum_{i:\overline{Q}_i \cap \gamma_j \neq \emptyset} \lambda(Q_i) = \frac{1}{Nr_j} \sum_{i:\overline{Q}_i \cap \gamma_j \neq \emptyset} d_j(S_i) \geq \frac{1}{Nr_j} r_j = \frac{1}{N}. \tag{3.46}$$

If this is not the case, then γ_j enters or exits $f^{-1}(A_j)$ through some peripheral disks Q_k, Q_l. Applying Corollary 3.7.4 to a subpath of γ_j that has its endpoints on S, and considering the contribution of $d_j(Q_k)$, $d_j(Q_l)$ we also obtain (3.46) in this case.

From now on, to fix our notation, we assume that the exceptional squares S_{i_1}, S_{i_2} that wish to exclude actually exist (if not, then one just has to ignore the indices i_1, i_2 in what follows). Summing in (3.46) over j we obtain

$$1 \leq \sum_{\substack{i:\overline{Q}_i \cap \beta_s \neq \emptyset \\ i \neq i_1, i_2}} \lambda(Q_i) \leq \sum_{\substack{i:\overline{Q}_i \cap \psi^{-1}(s) \neq \emptyset \\ i \neq i_1, i_2}} \lambda(Q_i).$$

Here ψ is as in Lemma 3.3.4. Observe that for all $i \in \mathbb{N}$ the functions $s \mapsto \chi_{\overline{Q}_i \cap \psi^{-1}(s)}$ are upper semi-continuous, thus measurable. We integrate over $s \in (0, \delta)$ and we obtain using Fubini's theorem:

$$\delta \leq \sum_{j=1}^{N} \frac{1}{Nr_j} \sum_{\substack{i:S_i \cap A_j \neq \emptyset \\ i \neq i_1, i_2}} d_j(S_i) \int_0^\delta \chi_{\overline{Q}_i \cap \psi^{-1}(s)} \, ds.$$

The fact that ψ is 1-Lipschitz yields

$$\delta \leq \sum_{j=1}^{N} \frac{1}{Nr_j} \sum_{\substack{i:S_i \cap A_j \neq \emptyset \\ i \neq i_1, i_2}} d_j(S_i) \, \mathrm{diam}(Q_i). \tag{3.47}$$

If we interchange the sums and apply the Cauchy–Schwarz inequality in the right hand side we have:

$$\frac{1}{N} \sum_{i \in \mathbb{N} \setminus \{i_1, i_2\}} \operatorname{diam}(Q_i) \sum_{j : S_i \cap A_j \neq \emptyset} \frac{d_j(S_i)}{r_j}$$

$$\leq \frac{1}{N} \left(\sum_{i \in \mathbb{N}} \operatorname{diam}(Q_i)^2 \right)^{1/2} \left[\sum_{i \in \mathbb{N} \setminus \{i_1, i_2\}} \left(\sum_{j : S_i \cap A_j \neq \emptyset} \frac{d_j(S_i)}{r_j} \right)^2 \right]^{1/2}.$$

Observe that the first sum is a finite constant $C^{1/2}$ by the quasiroundness assumption (3.1). In the second sum, note that each S_i, $i \neq i_1, i_2$, intersects only one annulus A_j so the inner sum actually is a sum over a single term. Combining these observations with (3.47), we have

$$\delta^2 N^2 \leq C \sum_{i \in \mathbb{N} \setminus \{i_1, i_2\}} \left(\sum_{j : S_i \cap A_j \neq \emptyset} \frac{d_j(S_i)^2}{r_j^2} \right) = C \sum_{j=1}^{N} \frac{1}{r_j^2} \sum_{\substack{i : S_i \cap A_j \neq \emptyset \\ i \neq i_1, i_2}} d_j(S_i)^2.$$

By (3.45), $d_j(S_i)^2 \leq C \mathcal{H}^2(S_i \cap A_j)$. Also,

$$\sum_{i : S_i \cap A_j \neq \emptyset} \mathcal{H}^2(S_i \cap A_j) \leq \mathcal{H}^2(A_j) = \pi \cdot (R_j^2 - r_j^2) = 3\pi r_j^2.$$

Hence,

$$\delta^2 N^2 \leq 3\pi C \sum_{j=1}^{N} \frac{1}{r_j^2} r_j^2 = 3\pi C N.$$

This is a contradiction as $N \to \infty$. $\square$

We have the following strong conclusion:

Corollary 3.8.2 *For all $i \in \mathbb{N}$ we have $\rho(Q_i) = \operatorname{osc}_{Q_i}(u) = \operatorname{osc}_{Q_i}(v) \neq 0$. In particular, no peripheral circle ∂Q_i is mapped under f to a point, and all squares S_i are non-degenerate.*

Remark 3.8.3 The non-vanishing of the "gradient" of a non-constant carpet-harmonic function is not clear in general. The difficulty in obtaining such a result in non-linear potential theory is also reflected by the fact that in dimension $n \geq 3$ it is not known whether the gradient of a non-constant p-harmonic function can vanish on an open set. This problem is a special case of the problem of unique continuation for the p-Laplace equation. See [20] for a partial result.

Remark 3.8.4 In [36] Schramm shows that finite triangulations of quadrilaterals can be transformed to square tilings of rectangles by a similar method. However, non-degeneracies cannot be avoided in his case, namely a vertex of the triangulation might correspond to a degenerate square under the correspondence. It is quite a surprise that, if one regards the carpet as an "infinite triangulation", these degeneracies disappear.

Corollary 3.8.5 *For each $t \in [0, 1]$ the level set $\alpha_t = \tilde{u}^{-1}(t)$ has empty interior.*

Proof Suppose that α_t has non-empty interior for some $t \in [0, 1]$. Then for some peripheral disk Q_i, $i \in \mathbb{N}$, the intersection $Q_i \cap \alpha_t$ contains an open set. Since $\tilde{u}$ is harmonic in Q_i (in the classical sense), it follows that $\tilde{u}$ is constant in $\overline{Q}_i$, and thus $\rho(Q_i) = 0$. This contradicts Corollary 3.8.2. □

Recall that if a level set α_t, $t \in \mathcal{T}$, intersects a peripheral circle ∂Q_i, then it intersects it in precisely two points. Using Lemma 3.8.1 we obtain a better description of the level sets α_t also for $t \notin \mathcal{T}$.

Lemma 3.8.6 *Fix $i \in \mathbb{N}$. For $t \in \{m_{Q_i}(u), M_{Q_i}(u)\}$ the level set α_t intersects ∂Q_i in a connected set, i.e., in an arc. Let β_1 be the arc corresponding to $t = m_{Q_i}(u)$ and β_3 be the arc corresponding to $t = M_{Q_i}(u)$. For all $t \in (m_{Q_i}(u), M_{Q_i}(u))$ the level set α_t intersects ∂Q_i in exactly two points, each of which is located on one of the two complementary arcs of β_1 and β_3 in ∂Q_i.*

Furthermore, $\Theta_1 = \alpha_0 \cap \partial\Omega$ and $\Theta_3 = \alpha_1 \cap \partial\Omega$. Finally, for all $t \in (0, 1)$ the intersection $\alpha_t \cap \partial\Omega$ contains two points, one on Θ_2 and one on Θ_4.

Proof Since $\mathrm{osc}_{Q_i}(u) > 0$ by Corollary 3.8.2, it follows that $m_{Q_i}(u) < M_{Q_i}(u)$. Suppose that for $t = m_{Q_i}(u)$ the set $\alpha_t \cap \partial Q_i$ has two distinct components $E_1, E_2 \subset \partial Q_i$. The sets E_1 and E_2 are closed arcs or points, they are disjoint, and $\partial Q_i \setminus (E_1 \cup E_2)$ has two components $F_1, F_2 \subset \partial Q_i$, which are open arcs. The function u is non-constant near the endpoints of F_j, on which it has the value t for $j = 1, 2$. In fact for $j = 1, 2$ we can find disjoint arcs $I_{j,k} \subset F_j$, $k = 1, 2$, on which u is non-constant and attains the value t at one of the endpoints of $I_{j,k}$, $k = 1, 2$. By continuity, it follows that $\bigcap_{j,k=1,2} u(I_{j,k})$ contains an interval of the form $(t, t + \varepsilon)$ for some $\varepsilon > 0$. The intermediate value theorem implies that we can find some $t' \in \mathcal{T} \cap (t, t + \varepsilon)$ such that $\alpha_{t'} \cap I_{j,k} \neq \emptyset$ for $j, k - 1, 2$. Hence, $\alpha_{t'} \cap \partial Q_i$ contains more than two points and this contradicts Corollary 3.5.8.

Let $\beta_1 \subset \partial Q_i$ be the closed arc that is the unique preimage in ∂Q_i under f of the left vertical side of ∂S_i. The same conclusion holds for $t = M_{Q_i}(u)$, and we consider the corresponding closed arc β_3.

Let $\beta_2, \beta_4 \subset \partial Q_i$ be the complementary closed arcs of the two extremal arcs, numbered in a counter-clockwise fashion, exactly as we numbered the sides of $\partial\Omega$. Since $f(\partial Q_i) = \partial S_i$ (by Corollary 3.7.2), the images of β_2 and β_4 are the bottom and top sides of the square ∂S_i; here it is not necessary that β_2 is mapped to the bottom side and β_4 is mapped to the top side, but it could be the other way around. It follows that v is constant on each of the arcs β_2, β_4.

By an application of the intermediate value theorem followed by the use of Corollary 3.5.8, one sees that for all $t \in (m_{Q_i}(u), M_{Q_i}(u))$ the intersection $\alpha_t \cap \partial Q_i$ must have two components, one in β_2, and one in β_4. Since u and v are constant on each component, these components must be singletons by Lemma 3.8.1.

The set $\alpha_0 \cap \partial\Omega$ contains Θ_1 and is contained in $\partial\Omega$. If there exists a point $x \in \alpha_0 \cap \Theta_2 \setminus \Theta_1$, then there has to be an entire arc $E \subset \alpha_0 \cap \Theta_2$. Otherwise, by the intermediate value theorem, there would exist level sets α_t, $t \in \mathcal{T}$, that intersect Θ_2 in at least two points, a contradiction to Lemma 3.5.6. However, v is identically equal to zero on E by Lemma 3.6.6, which implies that the continuum E is mapped to a point under f. This contradicts Lemma 3.8.1. The same argument applies with Θ_4 in the place of Θ_2 and yields that $\Theta_1 = \alpha_0 \cap \partial\Omega$. Similarly, $\Theta_3 = \alpha_1 \cap \partial\Omega$.

Finally, we already know from Proposition 3.5.1 that for all $t \in (0, 1)$ the intersection $\alpha_t \cap \partial\Omega$ has two components, one on Θ_2 and one on Θ_4. Again, by Lemma 3.8.1 these components have to be singletons. □

An immediate corollary is the following. We use notation from the proof of the preceding lemma.

Corollary 3.8.7 *For each peripheral disk Q_i, $i \in \mathbb{N}$, the arcs β_2 and β_4 are mapped injectively onto the bottom and top sides of ∂S_i, respectively. Furthermore, Θ_2 and Θ_4 are mapped injectively onto the bottom and top sides of $[0, 1] \times [0, D]$, respectively.*

Proof Recall that the winding number of $f\big|_{\partial\Omega}$ around every point of $(0, 1) \times (0, D)$ is $+1$ (see proof of Corollary 3.7.2). Moreover, we have

$$f(\overline{\Omega} \setminus Q_i) = [0, 1] \times [0, D] \setminus S_i,$$

by Corollary 3.7.2 and the fact that $f(Q_j) = S_j$ for all $j \in \mathbb{N}$; see the comments before Lemma 3.8.1. It follows by homotopy that the winding number of $f\big|_{\partial Q_i}$ around each point of S_i is $+1$. Hence, the arcs β_2 and β_4 of ∂Q_i must be mapped onto the bottom and top sides of ∂S_i, respectively.

Regarding the injectivity claim, note that f is injective when restricted to the "interior" of the arc β_2, since for each $t \in (m_{Q_i}(u), M_{Q_i}(u))$ the level set α_t intersects this arc at one point, by the previous lemma. By continuity, the endpoint $\beta_1 \cap \beta_2$ (strictly speaking, $\beta_1 \cap \beta_2$ is a singleton set containing one point) of β_2 has to be mapped to the bottom left corner of the square S_i, and the endpoint $\beta_3 \cap \beta_2$ of β_2 has to be mapped to the bottom right corner of S_i. This shows injectivity on all of β_2. Similarly, one shows the other claims about injectivity, based on the preceding lemma. □

We have completed our preparation to show the injectivity of f.

Lemma 3.8.8 *The map $f: S \to \mathcal{R}$ is injective.*

The proof of this lemma is slightly technical, so we first provide a sketch of the argument. Note that f is already injective when restricted to level sets $S \cap \alpha_t$ for $t \in \mathcal{T}$, by Corollary 3.6.7. In fact, v is increasing in some sense in these level sets.

Thus, the only possibility is that injectivity fails at some level set $\alpha_{t_0} \cap S$ for $t_0 \notin \mathcal{T}$. Suppose that $z = (t_0, s_0) \in \mathcal{R}$ has two preimages $x, y \in S \cap \alpha_{t_0}$. Then we show that for $t \in \mathcal{T}$ near t_0 there exist points $a_x = f^{-1}((t, s_x))$, $a_y = f^{-1}((t, s_y)) \in S \cap \alpha_t$ near x, y, respectively. In particular, $f(x)$ and $f(y)$ are near z by continuity. Using the fact that v is increasing on α_t, we show that there exists a path $\gamma \subset S$ connecting a_x, a_y that is mapped into a small neighborhood of z under f. In the limit, as $a_x \to x$ and $a_y \to y$ one obtains a continuum $E \subset S$ that connects x and y and is mapped to z. This will contradict Lemma 3.8.1.

Proof If $t \in \mathcal{T}$, then v is "increasing" on $\alpha_t \cap S$, in the sense of Corollary 3.6.7. Hence, f is injective on the set $U := \bigcup_{t \in \mathcal{T}} (\alpha_t \cap S)$.

Assume $z = (t_0, s_0) \in \mathcal{R}$ and $t_0 \notin \mathcal{T}$. Note that every point $x \in f^{-1}(z)$ can be approximated by Q_i with $\rho(Q_i) > 0$; see Corollary 3.8.2. Since $\rho(Q_i) > 0$, it follows that there exist levels $t \in \mathcal{T}$ with $\alpha_t \cap \partial Q_i \neq \emptyset$. Hence every point $x \in f^{-1}(z)$ can be approximated by points $a \in U$, and so U is dense in S. We will split in two cases. The ideas are similar but the technical details are slightly different. We recommend that the reader focus on Case 1 for a first reading of the proof.

Case 1 Assume first that $z = (0, s_0)$ or $(1, s_0)$ or z lies in an open vertical side (i.e., without the endpoints) of some square ∂S_{i_0}, $i_0 \in \mathbb{N}$. In any case, either every preimage $x \in f^{-1}(z)$ can only be approximated by points $a \in U$ with $u(a) < t_0$, or every preimage point can only be approximated by points $a \in U$ with $u(a) > t_0$. Indeed, this is clear if z lies in a vertical side of $[0, 1] \times [0, D]$ (i.e., $t_0 = 0$ or $t_0 = 1$), so suppose that $z = (t_0, s_0)$ lies in the open left vertical side of a square S_{i_0}, and x is a preimage of z. If $a_n \in U$ is a sequence converging to x, then $f(a_n) \in S$ converges to z by continuity. Since $a_n \in U$ and $t_0 \notin \mathcal{T}$, we necessarily have $u(a_n) \neq t_0$ for each $n \in \mathbb{N}$, and $f(a_n)$ cannot lie on the vertical line passing through z. Since z lies in the interior of the left vertical side of S_{i_0} and $f(a_n) \notin S_{i_0}$ for all $n \in \mathbb{N}$, it follows that $f(a_n)$ lies on the "left" of the vertical line $u = t_0$ for all sufficiently large n. Therefore, $u(a_n) < t_0$ for all sufficiently large n, as desired.

Suppose in what follows that every preimage point of z can only be approximated by points $a \in U$ with $u(a) < t_0$, and consider a preimage $x \in f^{-1}(z)$. Then, there exists a small ball $B(x, \delta)$ such that all points $a \in U$ that lie in the intersection $S \cap B(x, \delta)$ satisfy $u(a) < t_0$. Let $\delta' < \delta$ be so small that all peripheral disks intersecting $B(x, \delta')$, except for Q_{i_0} in the case $x \in \partial Q_{i_0}$, are contained in $B(x, \delta/2)$. Fix a point $a \in S \cap B(x, \delta')$ with $a \in U$ and $u(a) = t < t_0$. The level sets α_t, α_{t_0} determine a simply connected region A_{t,t_0} that contains all level sets $\alpha_{t'}$ for $t' \in (t, t_0)$, by Proposition 3.5.1. Since $B(x, \delta')$ meets both α_t and α_{t_0}, it follows that $\alpha_{t'}$ meets $B(x, \delta')$ for every $t' \in (t, t_0)$. The level sets $\alpha_{t'}$ cannot meet ∂Q_{i_0}, in the case $x \in \partial Q_{i_0}$ and $z \in \partial S_{i_0}$; this is because $f(\partial Q_{i_0}) = \partial S_{i_0}$. This implies that each point $b \in B(x, \delta') \cap \alpha_{t'}$ either lies in S, or it lies in a peripheral disk Q_i that is contained entirely in $B(x, \delta/2)$. In the second case, there exists a point in $\partial Q_i \cap \alpha_{t'} \subset S$ that is contained in $B(x, \delta)$. In any case, $S \cap B(x, \delta) \cap \alpha_{t'}$ is non-empty. Summarizing, for all $t < t_0$ sufficiently close to t_0 we have that the intersection $S \cap B(x, \delta) \cap \alpha_t$ is non-empty.

Suppose now that there exist two points $x, y \in f^{-1}(z)$, and consider a small $\delta > 0$ such that α_t intersects both sets $S \cap B(x, \delta)$, $S \cap B(y, \delta)$ for some $t < t_0$, $t \in \mathcal{T}$, sufficiently close to t_0. Let $a_x \in S \cap B(x, \delta) \cap \alpha_t$ and $a_y \in S \cap B(y, \delta) \cap \alpha_t$. By the continuity of f, the images $f(a_x) = (t, s_x)$, $f(a_y) = (t, s_y)$ will lie in a small ball $B(z, \varepsilon)$. Without loss of generality, we may suppose that $s_x < s_y$. Consider the path $\gamma \subset \alpha_t$ that connects a_x to a_y. The function v is increasing on α_t by Corollary 3.6.7, hence $f(\gamma \cap S) \subset \{t\} \times [s_x, s_y] \subset B(z, \varepsilon)$. We alter the path γ as follows. Whenever $\gamma \cap Q_i \neq \emptyset$ we replace this intersection with an arc $\beta \subset \partial Q_i$ joining the two points of $\gamma \cap \partial Q_i$. We call the resulting path $\tilde{\gamma}$ and note that $\tilde{\gamma} \subset S$. We claim that $f(\tilde{\gamma}) \subset B(z, 3\varepsilon)$. To see this, note that the image arcs $f(\beta)$ are contained in squares ∂S_i by Corollary 3.7.2, and the top and bottom sides of these squares intersect the ball $B(z, \varepsilon)$, namely at the endpoints of β which were also points of γ. Our claim is proved.

Finally, observe that as $\delta \to 0$ the path $\tilde{\gamma}$ subconverges to a non-trivial continuum $E \subset S$ containing x and y. Then the images $f(\tilde{\gamma})$ converge to the point z, so by continuity $f(E) = z$. This contradicts Lemma 3.8.1.

Case 2 Assume that $z = (t_0, s_0)$ does not lie on an open vertical side of any square ∂S_i, $i \in \mathbb{N}$, or on a vertical side of the rectangle $[0, 1] \times [0, D]$ (so $0 < t_0 < 1$). We claim that there exists a *distinguished* preimage x of z that can be approximated by points $a, b \in [0, 1] \times [0, D]$ with $\tilde{u}(a) < t_0$ and $\tilde{u}(b) > t_0$.

To prove that, we will show that there exists a continuum $C \subset \alpha_{t_0}$ connecting Θ_2 and Θ_4 such that every point $x \in C$ can be approximated by points a, b with $\tilde{u}(a) < t_0$ and $\tilde{u}(b) > t_0$. The function $\tilde{v}$ is continuous on C and attains all values between 0 and D. Hence, there exists some $x \in C$ with $\tilde{v}(x) = s_0$. In fact, $x \in C \cap S \cap f^{-1}(z)$, because $f(Q_i) = S_i$ for all $i \in \mathbb{N}$; see comments before Lemma 3.8.1.

The existence of C will be justified with the following lemma about planar topology, which reflects the *unicoherence* of the plane:

Lemma 3.8.9 ([45, Theorem 5.28a, p. 65]) *If A, B are planar continua neither of which separates the plane, then $A \cup B$ does not separate the plane if and only if $A \cap B$ is a continuum.*

Let $V_1 = \tilde{u}^{-1}((0, t_0)) = A_{0,t_0}$ and $V_3 = \tilde{u}^{-1}((t_0, 1)) = A_{t_0,1}$. Since all level sets α_t have empty interior by Corollary 3.8.5, Proposition 3.5.1 implies that the closures $\overline{V}_1, \overline{V}_3$ are continua that do not separate the plane. Moreover, $\overline{V}_1 \cup \overline{V}_3 = \overline{\Omega}$ so $\overline{V}_1 \cup \overline{V}_3$ does not separate the plane. Indeed, a point $w \in \overline{\Omega} \setminus \overline{V}_1 \cup \overline{V}_3$ would necessarily satisfy $u(w) = t_0$ and would have a rel. open neighborhood W in $\overline{\Omega} \setminus \overline{V}_1 \cup \overline{V}_3$. This would imply that α_{t_0} has non-empty interior, a contradiction. By Lemma 3.8.9 we conclude that $\overline{V}_1 \cap \overline{V}_3 \subset \alpha_{t_0}$ is a continuum. Moreover, this continuum has to connect $\alpha_{t_0} \cap \Theta_2$ to the point $\alpha_{t_0} \cap \Theta_4$. This is because $\alpha_{t_0} \cap \Theta_2$, $\alpha_{t_0} \cap \Theta_4$ are singletons by Lemma 3.8.6, and they lie in $\overline{V}_1 \cap \overline{V}_3$ since their complement in Θ_2, Θ_4, respectively, consists of two arcs, one contained in V_1 and one in V_3. Summarizing, $C := \overline{V}_1 \cap \overline{V}_3$ is the desired continuum connecting Θ_2 to Θ_4.

Our next claim is that for every small ball $B(x, \delta)$, where $x \in C$ is a preimage of z, the set $S \cap B(x, \delta) \cap \alpha_t$ is non-empty for all $t < t_0$ sufficiently close to t_0, and for all $t > t_0$ sufficiently close to t_0; this will be the crucial property that we need for the distinguished preimage x of z. To prove this, we consider three cases: $x \in S^\circ$, $x \in \partial Q_i$ for some $i \in \mathbb{N}$, and $x \in \partial \Omega$.

First, assume that $x \in S^\circ$. We fix a small ball $B(x, \delta)$ and a point $a \in B(x, \delta)$ with $\tilde{u}(a) = t < t_0$; such a point exists by the properties of $x \in C$. The level sets α_t, α_{t_0} define a simply connected region A_{t,t_0}, and all level sets $\alpha_{t'}$ for t' between t and t_0 lie in this region and have to intersect $B(x, \delta)$; see Proposition 3.5.1. If $x \in S^\circ$ then $S \cap B(x, \delta) \cap \alpha_{t'} \neq \emptyset$, since the peripheral disks that might be intersected by sets α'_t for $t' \in (t, t_0)$ are arbitrarily small; see also Case 1. The same holds for level sets α_t, $t > t_0$, sufficiently close to t_0.

We now prove the claim in the case $x \in \partial Q_{i_0}$ for some $i_0 \in \mathbb{N}$. Recall by Corollary 3.8.7 that ∂Q_{i_0} is partitioned into four arcs $\beta_1, \ldots, \beta_4$, such that β_2, β_4 are mapped injectively onto the top an bottom sides of ∂S_{i_0}, respectively, and β_1, β_3 are mapped injectively onto the left and right sides of ∂Q_i, respectively, by the Case 1. In particular, $f|_{\partial Q_i}$ is injective and $f|_{\partial Q_i}^{-1}$ is defined and is continuous on ∂S_i. The only possibility here is that $x \in \beta_2 \cup \beta_4$ (recall the assumption of Case 2). If x is an interior point of one of the arcs β_2, β_4, then $\partial Q_{i_0} \cap S \cap B(x, \delta) \cap \alpha_t \neq \emptyset$ for all t near t_0 by the continuity of $f|_{\partial Q_i}^{-1}$. If x is a "corner" point lying, e.g., on $\beta_2 \cap \beta_1$ then we can approximate x by points $a \in \beta_2$ thus satisfying $u(a) > t_0$, and we can approximate x by points $b \in \beta_1$ with $u(b) = t_0$. In the latter case, $f(b)$ is contained in the left open vertical side of ∂S_{i_0}. Arguing as in Case 1, for a small $\delta' > 0$ we have $S \cap B(b, \delta') \cap \alpha_t \neq \emptyset$ for all $t < t_0$ sufficiently close to t_0. If b is sufficiently close to x then $B(b, \delta') \subset B(x, \delta)$, so $S \cap B(x, \delta) \cap \alpha_t \neq \emptyset$, as desired.

Finally, if $x \in \partial \Omega$, then it has to lie in the interior of the arcs Θ_2 or Θ_4. The map f is injective on these arcs by Corollary 3.8.7. Hence it is easy to see that $S \cap B(x, \delta) \cap \alpha_t \neq \emptyset$ for all t near t_0, and in fact the intersection contains points of Θ_2 or Θ_4.

The preimage x is the distinguished preimage of z that is "visible" from both sides $u < t_0$ and $u > t_0$. Now, assume that there exists another preimage $y \in f^{-1}(z)$. Since U is dense in S, there exist near y points $a \in \alpha_t \cap S$, $t \in \mathcal{T}$, with $u(a) < t_0$ or $u(a) > t_0$. Without loss of generality, we assume that arbitrarily close to y we can find such points with $u(a) = t < t_0$. Then for a small $\delta > 0$ the intersection $S \cap B(y, \delta) \cap \alpha_t$ is non-empty for $t < t_0$, $t \in \mathcal{T}$, arbitrarily close to t_0. Now, we argue exactly as in the last part of Case 1. Consider a level set α_t, $t \in \mathcal{T}$, that intersects both sets $S \cap B(x, \delta)$ and $S \cap B(y, \delta)$ and let $\gamma \subset \alpha_t$ be a subpath that connects the balls $B(x, \delta)$, $B(y, \delta)$. The image $f(\gamma \cap S)$ is contained in a small ball $B(z, \varepsilon)$. We modify the path γ to obtain a path $\tilde{\gamma} \subset S$ that connects the balls $B(x, \delta)$, $B(y, \delta)$, and such that $f(\tilde{\gamma})$ lies in a slightly larger ball $B(z, 3\varepsilon)$. As $\delta \to 0$ the path $\tilde{\gamma}$ subconverges to a continuum in S that connects x and y. The images $f(\tilde{\gamma})$ converge to z, so Lemma 3.8.1 is again contradicted. $\square$

Remark 3.8.10 Since $f\big|_{\partial Q_i}$ is a homeomorphism and $\tilde{u}$, $\tilde{v}$ are harmonic on Q_i, we can use the Radó–Kneser–Choquet theorem [16, p. 29] to conclude that $\tilde{f}\big|_{\overline{Q}_i}$ is a homeomorphism onto $\overline{S}_i$. Thus, $f: \overline{\Omega} \to [0, 1] \times [0, D]$ is a homeomorphism. Furthermore, the set $\mathcal{R} = f(S)$ is a Sierpiński carpet, as defined in the Introduction of the current chapter (Sect. 3.1), with $\mathcal{H}^2(\mathcal{R}) = 0$ and the basic assumptions of quasiroundness (3.1) and Ahlfors 2-regularity (3.2) are trivially satisfied for the peripheral disks of $\mathcal{R}$, which are squares.

In fact, the harmonicity of the extension inside each Q_i is not needed anymore, so one could consider arbitrary homeomorphic extensions $\tilde{f}: \overline{Q}_i \to \overline{S}_i$ given, for example, by the Schönflies theorem. Recall that $Q_0 = \mathbb{C} \setminus \overline{\Omega}$ and $S_0 = \mathbb{C} \setminus [0, 1] \times [0, D]$. We also extend $f\big|_{\partial Q_0}$ to a homeomorphism from $\overline{Q}_0$ onto $\overline{S}_0$. Pasting together these homeomorphisms we obtain a homeomorphic extension $\tilde{f}: \mathbb{C} \to \mathbb{C}$ of f; see e.g. [7, Lemma 5.5].

3.9 Regularity of f and f^{-1}

In this section we prove the main result, Theorem 3.1.1, which will follow from Proposition 3.9.4 and Proposition 3.9.5. We consider an (arbitrary) homeomorphic extension of $f: S \to \mathcal{R}$ to a map $f: \mathbb{C} \to \mathbb{C}$, as in Remark 3.8.10. As discussed in this remark, the set $\mathcal{R}$ is a Sierpiński carpet, as defined in the Introduction of the current chapter (Sect. 3.1).

Recall that by Proposition 3.7.3 we have the upper gradient inequality

$$|f(x) - f(y)| \leq \sqrt{2} \sum_{i:\overline{Q}_i \cap \beta \neq \emptyset} \rho(Q_i) = \sum_{i:\overline{Q}_i \cap \beta \neq \emptyset} \text{diam}(S_i) \tag{3.48}$$

for points x, $y \in \overline{\beta} \cap S$ and all open paths β, which are subpaths of paths $\gamma \subset \Omega$ lying outside an exceptional family of 2-modulus zero. We will use this to show that f preserves, in some sense, carpet modulus. In fact, for technical reasons, we will introduce a slightly different notion of carpet modulus here, which was also used in the statement of Theorem 3.1.1:

Definition 3.9.1 Let Γ be a family of paths in $\mathbb{C}$. A sequence of non-negative numbers $\{\lambda(Q_i)\}_{i \in \mathbb{N} \cup \{0\}}$ is *admissible for the weak carpet modulus* $\overline{\text{mod}}(\Gamma)$ if there exists an exceptional path family Γ_0 with $\text{mod}_2(\Gamma_0) = 0$ such that for all $\gamma \in \Gamma \setminus \Gamma_0$ we have

$$\sum_{i:\overline{Q}_i \cap \gamma \neq \emptyset} \lambda(Q_i) \geq 1.$$

We define $\overline{\text{mod}}(\Gamma) = \inf_\lambda \sum_{i \in \mathbb{N} \cup \{0\}} \lambda(Q_i)^2$ where the infimum is taken over all admissible weights λ.

We would like to point out at this point that $Q_0 = \mathbb{C} \setminus \overline{\Omega}$, and we *do* include $\lambda(Q_0)$ in the above sums whenever $\overline{Q_0} \cap \gamma \neq \emptyset$, in contrast to the previous sections, in which all paths were contained in Ω.

First we show a preliminary lemma, in the same spirit as Corollary 3.7.4.

Lemma 3.9.2 *There exists a path family Γ_1 in $\mathbb{C}$ with $\mathrm{mod}_2(\Gamma_1) = 0$ such that for every path $\gamma \subset \mathbb{C}$, $\gamma \notin \Gamma_1$, the following holds:*

For any two points $x, y \in \overline{\gamma}$, there exists an open path $\tilde{\gamma} \subset \mathbb{C}$ joining x and y, such that $\{i \in \mathbb{N} \cup \{0\} : \overline{Q}_i \cap \tilde{\gamma} \neq \emptyset\} \subset \{i \in \mathbb{N} \cup \{0\} : \overline{Q}_i \cap \gamma \neq \emptyset\}$ and $\mathcal{H}^1(f(\tilde{\gamma} \cap S)) = 0$.

The curve $\tilde{\gamma}$ is not necessarily a subcurve of γ, but it intersects no more (closed) peripheral disks than γ does.

Remark 3.9.3 An important ingredient in the proof will be the following. Although the upper gradient inequality (3.48) holds—a priori—only for subpaths of paths γ *contained* in Ω and lying outside an exceptional family Γ_0 with $\mathrm{mod}_2(\Gamma_0)$, it turns out that this can be extended to paths in $\mathbb{C}$. Namely, the family of paths in $\mathbb{C}$ that have a subpath $\gamma \subset \Omega$ for which the upper gradient inequality fails has 2-modulus zero. This implies that (3.48) holds for all open subpaths $\beta \subset \Omega$ of paths $\gamma \subset \mathbb{C}$ that lie outside an exceptional family Γ_0 of 2-modulus zero.

Proof Let $\gamma \subset \mathbb{C}$ be a path such that:

1. (3.48) holds along all of its subpaths that are contained in Ω,
2. $\mathcal{H}^1(\gamma \cap S) = 0$, and
3.

$$\sqrt{2} \sum_{\substack{i: \overline{Q}_i \cap \gamma \neq \emptyset \\ i \in \mathbb{N} \setminus \{0\}}} \rho(Q_i) = \sum_{\substack{i: \overline{S}_i \cap f(\gamma) \neq \emptyset \\ i \in \mathbb{N} \setminus \{0\}}} \mathrm{diam}(S_i) < \infty.$$

This is satisfied by all paths outside an exceptional family Γ_1 with $\mathrm{mod}_2(\Gamma_1) = 0$. Indeed, (1) holds for mod_2-a.e. path by the preceding remark, and (2) holds for mod_2-a.e. path since $\mathcal{H}^2(S) = 0$. The third statement also holds for all paths outside a family of conformal modulus zero by Remark 3.7.6. By the subadditivity of conformal modulus, all three conditions are simultaneously met by all paths outside a family of conformal modulus zero, as desired.

Let $x, y \in \overline{\gamma}$. Note that if x, y lie on the same peripheral disk $\overline{Q}_{i_0}$, $i_0 \in \mathbb{N} \cup \{0\}$, and γ intersects $\overline{Q}_{i_0}$, then there is nothing to show, since we can just connect x, y with an open path $\tilde{\gamma} \subset Q_{i_0}$, and this will trivially have the desired properties. Hence, we may assume that this is not the case. By replacing γ with an open subpath we may suppose that x, y are the endpoints of $\overline{\gamma}$. Assume that γ is parametrized as it runs from x to y. If $x \in \overline{Q}_{i_x}$ for some $i_x \in \mathbb{N} \cup \{0\}$ and $\overline{Q}_{i_x} \cap \gamma \neq \emptyset$ then we can replace x with the last exit point $x_0 \in \partial Q_{i_x}$ of γ from $\overline{Q}_{i_x}$, and obtain similarly a point $y_0 \in \partial Q_{i_y}$, which is the first entry point of γ (after x_0) in $\overline{Q}_{i_y}$, in the case $y \in \partial Q_{i_y}$ for some $i_y \in \mathbb{N} \cup \{0\}$ (cf. the discussion on accessible points following

Definition 3.4.3). If we can find a path $\tilde{\gamma}$ joining x_0, y_0 with the desired properties, then we can concatenate it with arcs inside Q_{i_x}, Q_{i_y}, and these do not contribute to the Hausdorff 1-measure of $f(\tilde{\gamma} \cap S)$, so the conclusion holds for the concatenation. Hence, we assume that $x, y \in \overline{\gamma} \cap S$ and γ does not intersect the closures of the peripheral disks that possibly contain x, y in their boundary.

Another reduction we make is to assume that $\gamma \subset \Omega$. Indeed, by the previous paragraph we may assume that none of x, y lies in Q_0 and that γ does not intersect $\overline{Q}_0$ in the case $x \in \partial Q_0$ or $y \in \partial Q_0$. If γ intersects $\overline{Q}_0$ (thus $x, y \in \Omega$), we let x_0 be the first entry point of γ in $\overline{Q}_0$, and y_0 be the last exit point from $\overline{Q}_0$. We consider the open subpaths $\gamma_x \subset \Omega$ from x to x_0 and $\gamma_y \subset \Omega$ from y to y_0. If the statement of the lemma is true for γ_x and γ_y, then there exist paths $\tilde{\gamma}_x$, $\tilde{\gamma}_y$ joining x to x_0, y to y_0, respectively, such that they do not intersect more closed peripheral disks than γ does, and such that their image has length zero in the carpet $\mathcal{R}$. Concatenating $\tilde{\gamma}_x$, $\tilde{\gamma}_y$ with a path inside Q_0 that joins x_0 to y_0 provides the desired path $\tilde{\gamma}$.

Assuming now that $\gamma \subset \Omega$ and γ does not intersect the closed peripheral disks possibly containing x, y in their boundaries, we will construct the path $\tilde{\gamma}$ through some iteration procedure. We fix $\varepsilon > 0$, and a finite index set $J_1 \subset \mathbb{N}$ such that

$$\sum_{\substack{i:\overline{Q}_i \cap \gamma \neq \emptyset \\ i \in \mathbb{N} \setminus J_1}} \operatorname{diam}(S_i) < \varepsilon. \tag{3.49}$$

Using Lemma 3.3.8, we may obtain open subpaths $\gamma_1, \ldots, \gamma_m \subset \Omega$ of γ that intersect disjoint sets of peripheral disks $\overline{Q}_i$, $i \notin J_1$, such that the upper gradient inequality of f holds along each of them and they have the following property: the path γ_1 starts at $x_1 = x$, the path γ_m terminates at $y_m = y$ and in general the path γ_k has endpoints $x_k, y_k \in S$ such that for $1 \leq k \leq m - 1$ we either have

- $y_k = x_{k+1}$, or
- $y_k, x_{k+1} \in \partial Q_{j_k}$ for some $j_k \in \mathbb{N}$. The peripheral disks $\overline{Q}_{j_k}$ are distinct and they are intersected by γ.

We denote by $I \subset \{1, \ldots, m\}$ the set of indices k for which the second alternative holds. By shrinking the open paths γ_k (or even discarding some of them), we may further assume that each path γ_k does not intersect the peripheral disks that possibly contain the endpoints of γ_k in their boundary. See also the proof of Corollary 3.7.4.

For each $k \in I$ we consider a path $\alpha_k \subset \Omega$ such that $f(\alpha_k)$ is a line segment inside $S_{j_k} = f(Q_{j_k})$ that connects the endpoints of $f(\gamma_k)$ and $f(\gamma_{k+1})$. Concatenating all the paths α_k, γ_k we obtain a path β_1 such that $f(\beta_1)$ connects $f(x)$ to $f(y)$ and intersects fewer peripheral disks than $f(\gamma)$. We estimate $\mathcal{H}^1(f(\beta_1 \cap S))$ as follows: for each $k \in \{1, \ldots, m\}$ the image $f(\overline{\gamma}_k \cap S)$ can be covered by a ball B_k of radius $r_k = \sum_{i:\overline{Q}_i \cap \gamma_k \neq \emptyset} \operatorname{diam}(S_i)$, because by the upper gradient inequality (3.48) we have

$$\sup_{z,w \in \overline{\gamma}_k \cap S} |f(z) - f(w)| \leq \sum_{i:\overline{Q}_i \cap \gamma_k \neq \emptyset} \operatorname{diam}(S_i).$$

By construction, $\gamma_1, \ldots, \gamma_m$ intersect disjoint sets of peripheral disks. Since $f(\beta_1 \cap S) = \bigcup_{k \in I} f(\overline{\gamma}_k \cap S)$, we have that

$$\mathcal{H}_{2\varepsilon}^1(f(\beta_1 \cap S)) \le 2 \sum_{k=1}^{m} r_k \le 2 \sum_{k=1}^{m} \sum_{i:\overline{Q}_i \cap \gamma_k \ne \emptyset} \mathrm{diam}(S_i) < 2\varepsilon$$

by (3.49). Here it is crucial that the paths γ_k intersect disjoint sets of peripheral disks.

Now, we iterate this procedure with ε replaced by $\varepsilon/2$. We consider a finite index set $J_2 \supset J_1$ such that

$$\sum_{\substack{i:\overline{Q}_i \cap \gamma \ne \emptyset \\ i \in \mathbb{N} \setminus J_2}} \mathrm{diam}(S_i) < \varepsilon/2$$

and we modify each of the paths $\gamma_1, \ldots, \gamma_m$ from the first step, according to the previous procedure. The paths α_k from the first step remain unchanged, but new arcs $\alpha_l \subset Q_l$ will be added from the second step, such that $\overline{Q}_l \cap \gamma_k \ne \emptyset$ for some $k \in \{1, \ldots, m\}$, and thus

$$\mathrm{diam}(f(\alpha_l)) \le \mathrm{diam}(S_l) \le \sum_{i:\overline{Q}_i \cap \gamma_k \ne \emptyset} \mathrm{diam}(S_i) = r_k.$$

This implies that $f(\alpha_l) \subset S_l \subset 2B_k$. Hence, the new path $f(\beta_2)$ that we obtain as a result of concatenations connects $f(x)$ to $f(y)$ and stays close to the path $f(\beta_1)$. In fact, we can achieve that $f(\beta_2 \cap S)$ is covered by balls B_l of radius r_l, such that $2B_l$ is contained in one of the balls $2B_k$, and

$$\mathcal{H}_\varepsilon^1(f(\beta_2 \cap S)) \le 2 \sum_l r_l \le \varepsilon.$$

This can be achieved by noting that $f(\beta_2 \cap S) \subset f(\beta_1 \cap S) \subset \bigcup_k B_k$, and choosing an even larger set $J_2 \supset J_1$, so that

$$r_l \le \sum_{\substack{i:\overline{Q}_i \cap \gamma \ne \emptyset \\ i \in \mathbb{N} \setminus J_2}} \mathrm{diam}(S_i) < \min_{k \in \{1, \ldots, m\}} r_k/2.$$

In the n-th step we have a path $f(\beta_n)$ connecting $f(x), f(y)$ such that the set $f(\beta_n \cap S)$ admits a cover by balls $\{2B_l^n\}_l$ that is decreasing in n, and whose radii sum does not exceed $\varepsilon/2^{n-2}$. Moreover, by construction, if $f(\beta_n) \cap S_i \ne \emptyset$ for a square $i \in \mathbb{N}$, then $f(\beta_m) \cap S_i$ is a fixed line segment for all $m \ge n$.

The curves $f(\beta_n)$ subconverge to a continuum $f(\beta)$ in the Hausdorff sense that connects $f(x)$ to $f(y)$. The set $f(\beta \cap S)$ is covered by $\{2B_l^n\}_l$ for all n and thus $\mathcal{H}^1(f(\beta \cap S)) = 0$. The set β can only intersect peripheral disks that can be approximated by points intersected by the paths β_n, and thus by the path γ. Hence, β cannot intersect more peripheral disks than γ does.

We claim that $f(\beta)$ is locally connected, and thus it contains a path connecting $f(x)$, $f(y)$; see the proof of Lemma 3.5.13. The argument is similar to the proof of Lemma 3.5.14. If $f(\beta)$ is not locally connected, there exists an open set U and $\varepsilon > 0$ such that $U \cap f(\beta)$ contains infinitely many components C_n, $n \in \mathbb{N}$, of diameter at least ε. By passing to a subsequence, we may assume that the continua $\overline{C}_n$ subconverge in the Hausdorff sense to a continuum $C \subset f(\beta)$ with $\mathrm{diam}(C) \geq \varepsilon$. We claim that $C \subset \mathcal{R}$, hence $\mathcal{H}^1(C) \leq \mathcal{H}^1(f(\beta \cap S)) = 0$, which is a contradiction.

If $C \cap S_{i_0} \neq \emptyset$ for some $i_0 \in \mathbb{N}$, then by shrinking C and C_n we may assume that $C, C_n \subset\subset S_{i_0}$ for all $n \in \mathbb{N}$. In particular, $f(\beta)$ has infinite length inside S_{i_0}. However, by the construction of the paths $f(\beta_n)$, the intersection $f(\beta) \cap S_{i_0}$ is either empty or it is one line segment. This is a contradiction. □

Using the lemma we show the following.

Proposition 3.9.4 *Let Γ be a family of paths in $\mathbb{C}$ joining two continua $E, F \subset S$, but avoiding finitely many peripheral disks $\overline{Q}_i$, $i \in I_0$, where I_0 is a finite (possibly empty) subset of $\mathbb{N} \cup \{0\}$. We have*

$$\overline{\mathrm{mod}}(\Gamma) \leq \mathrm{mod}(f(\Gamma)).$$

Here, as before, the left hand side is weak carpet modulus with respect to the carpet S, and the right hand side is carpet modulus with respect to the carpet $\mathcal{R}$.

Proof Consider a weight $\{\lambda'(S_i)\}_{i \in \mathbb{N} \cup \{0\}}$ that is admissible for $\mathrm{mod}(f(\Gamma))$, i.e.,

$$\sum_{i: S_i \cap \gamma' \neq \emptyset} \lambda'(S_i) \geq 1 \tag{3.50}$$

for every path $\gamma' = f(\gamma) \in f(\Gamma)$ with $\mathcal{H}^1(\gamma' \cap \mathcal{R}) = 0$. Define $\lambda(Q_i) = \lambda'(S_i)$, $i \in \mathbb{N} \cup \{0\}$, and let $\Gamma_0 = \{\gamma \in \Gamma : \sum_{i: \overline{Q}_i \cap \gamma \neq \emptyset} \lambda(Q_i) < 1\}$. We wish to show that $\mathrm{mod}_2(\Gamma_0) = 0$. If this is the case, then $\{\lambda(Q_i)\}_{i \in \mathbb{N}}$ is clearly admissible for $\overline{\mathrm{mod}}(\Gamma)$, thus

$$\overline{\mathrm{mod}}(\Gamma) \leq \sum_{i \in \mathbb{N} \cup \{0\}} \lambda(Q_i)^2 = \sum_{i \in \mathbb{N} \cup \{0\}} \lambda'(S_i)^2.$$

Infimizing over λ we obtain the desired $\overline{\mathrm{mod}}(\Gamma) \leq \mathrm{mod}(f(\Gamma))$.

Now we show our claim. If $\gamma \in \Gamma_0 \setminus \Gamma_1$, where Γ_1 is as in Lemma 3.9.2, then there exists a path $\tilde{\gamma}$ given by the lemma that connects points $x \in E$ and $y \in F$. The path $\tilde{\gamma}$ intersects fewer peripheral disks than γ, so

$$\sum_{i: S_i \cap f(\tilde{\gamma}) \neq \emptyset} \lambda'(S_i) = \sum_{i: Q_i \cap \tilde{\gamma} \neq \emptyset} \lambda(Q_i) \leq \sum_{i: \overline{Q}_i \cap \gamma \neq \emptyset} \lambda(Q_i) < 1.$$

Also, $\mathcal{H}^1(f(\tilde{\gamma} \cap S)) = \mathcal{H}^1(f(\tilde{\gamma}) \cap \mathcal{R}) = 0$, $\tilde{\gamma}$ still joins the continua E, F, and avoids $\bigcup_{i \in I_0} Q_i$, so $\tilde{\gamma} \in \Gamma$. This contradicts (3.50), hence $\Gamma_0 \setminus \Gamma_1 = \emptyset$, and $\Gamma_0 \subset \Gamma_1$. We thus have $\mathrm{mod}_2(\Gamma_0) \leq \mathrm{mod}_2(\Gamma_1) = 0$.   □

We also wish to show an analog of Proposition 3.9.4 for $g = f^{-1}$:

Proposition 3.9.5 *Let* Γ *be the family of paths in* $\mathbb{C}$ *joining two continua* E, $F \subset \mathcal{R}$, *but avoiding finitely many squares* $\overline{S}_i$, $i \in I_0$, *where* I_0 *is a finite (possibly empty) subset of* $\mathbb{N} \cup \{0\}$. *We have*

$$\overline{\mathrm{mod}}(\Gamma) \leq \mathrm{mod}(g(\Gamma)).$$

The proof of Proposition 3.9.4 applies without change, if we can establish an analog of Lemma 3.9.2 for g. In the proof of the latter we only used the fact that f is a homeomorphism, together with (3.48). Hence, in order to obtain Proposition 3.9.5, it suffices to show that

$$|g(x) - g(y)| \leq \sum_{i:\overline{S}_i \cap \gamma \neq \emptyset} \mathrm{diam}(Q_i) \tag{3.51}$$

for all paths $\gamma \subset (0, 1) \times (0, D)$ outside an exceptional family of 2-modulus zero, and points x, $y \in \overline{\gamma} \cap \mathcal{R}$. We will show this in Lemma 3.9.8 after we have established two auxiliary results.

Recall the notation $\beta_s \subset \psi^{-1}(s) \cap ((0, 1) \times (0, D))$, for a.e. $s \in (0, \delta)$, that is used to denote the perturbations of a given path β connecting two non-trivial continua E, $F \subset [0, 1] \times [0, D]$, as in Proposition 3.3.4. Here $\psi(x) = \mathrm{dist}(x, \beta)$.

Lemma 3.9.6 *Fix a path* β *as above. For a.e.* $s \in (0, \delta)$ *we have*

$$\mathcal{H}^1(g(\psi^{-1}(s) \cap \mathcal{R})) = 0.$$

We postpone the proof for the moment. Using this we have a preliminary version of (3.51):

Lemma 3.9.7 *For a.e.* $s \in (0, \delta)$ *and for all* x, $y \in \overline{\beta}_s \cap \mathcal{R}$ *we have*

$$|g(x) - g(y)| \leq \sum_{i:\overline{S}_i \cap \beta_s \neq \emptyset} \mathrm{diam}(Q_i).$$

Proof Consider $s \in (0, \delta)$ such that the conclusion of Lemma 3.9.6 holds. Without loss of generality, we assume that $\sum_{i:\overline{S}_i \cap \beta_s \neq \emptyset} \mathrm{diam}(Q_i) < \infty$. Let $\varepsilon > 0$ and consider a cover of $g(\beta_s \cap \mathcal{R}) \subset g(\psi^{-1}(s) \cap \mathcal{R})$ by finitely many small balls B_j of radius r_j such that $\sum_j r_j < \varepsilon$. The union of $\bigcup_j B_j$ and $\bigcup_{i:\overline{S}_i \cap \beta_s \neq \emptyset} Q_i$ covers $g(\beta_s)$ so we can find a finite subcover. Traveling along this subcover from $g(x)$ to

$g(y)$ (as a finite chain) we obtain

$$|g(x) - g(y)| \le \sum_j 2r_j + \sum_{i:\overline{S}_i \cap \beta_s \ne \emptyset} \text{diam}(Q_i) \le 2\varepsilon + \sum_{i:\overline{S}_i \cap \beta_s \ne \emptyset} \text{diam}(Q_i).$$

The conclusion follows. $\square$

Now, we establish the following lemma.

Lemma 3.9.8 *There exists an exceptional family of paths Γ_0 with $\text{mod}_2(\Gamma_0) = 0$ such that for every path $\gamma \subset (0, 1) \times (0, D)$ with $\gamma \notin \Gamma_0$ and every open subpath β of γ we have*

$$|g(x) - g(y)| \le \sum_{i:\overline{S}_i \cap \beta \ne \emptyset} \text{diam}(Q_i) \tag{3.52}$$

for all points $x, y \in \overline{\beta} \cap \mathcal{R}$.

Proof Note that Lemma 3.6.9 is also valid for the carpet $\mathcal{R}$, since it satisfies the basic assumptions (3.1) and (3.2) and has area zero (see Corollary 3.7.2). We fix a non-exceptional path $\gamma \subset (0, 1) \times (0, D)$ so that the conclusions of Lemma 3.6.9 hold for all subpaths β of γ having endpoints in $\mathcal{R}^\circ$, and $\mathcal{H}^1(\gamma \cap \mathcal{R}) = 0$.

By continuity and the usual reduction to accessible points, it suffices to prove the main claim (3.52) for a subpath β of γ and its endpoints x, y, with the further assumption that they lie in $\mathcal{R}^\circ$ (i.e. they do not lie on any square or on the boundary rectangle); see the comments in the beginning of the proof of Lemma 3.6.8. Consider two non-trivial disjoint continua $E, F \subset \mathcal{R}^\circ$ such that $x \in E$ and $y \in F$; the existence of such continua is implied by Lemma (b). Let $\psi(x) = \text{dist}(x, \beta)$ and consider the perturbations $\beta_s \subset \psi^{-1}(s)$ of β that connect points $x_s \in E$ to $y_s \in F$, for a.e. $s \in (0, \delta)$. If δ is sufficiently small, then $\psi^{-1}(s) \subset (0, 1) \times (0, D)$ for all $s \in (0, \delta)$.

For fixed $\varepsilon > 0$ we choose an even smaller $\delta > 0$ so that x_s and y_s are close to x and y, respectively, and

$$|g(x) - g(y)| \le |g(x_s) - g(y_s)| + \varepsilon$$

$$\le \sum_{i:\overline{S}_i \cap \psi^{-1}(s) \ne \emptyset} \text{diam}(Q_i) + \varepsilon$$

for a.e. $s \in (0, \delta)$, by the continuity of g and Lemma 3.9.7. The right hand side is a measurable function of s. Thus, averaging over $s \in (0, \delta)$ and using Fubini's theorem we obtain

$$|g(x) - g(y)| \le \frac{1}{\delta} \sum_{i \in \mathbb{N}} \text{diam}(Q_i) \int_0^\delta \chi_{\overline{S}_i \cap \psi^{-1}(s)} \, ds + \varepsilon$$

$$\le \frac{1}{\delta} \sum_{i \in \mathbb{N}} \text{diam}(Q_i) \mathcal{H}^1(\{s \in (0, \delta) : \overline{S}_i \cap N_s(\beta) \ne \emptyset\}) + \varepsilon$$

$$\leq \sum_{i:\overline{S}_i \cap \beta \neq \emptyset} \mathrm{diam}(Q_i)$$

$$+ \frac{1}{\delta} \sum_{\substack{i:\overline{S}_i \cap \beta = \emptyset \\ S_i \cap N_\delta(\beta) \neq \emptyset}} \mathrm{diam}(Q_i) \cdot \min\{\mathrm{diam}(S_i), \delta\} + \varepsilon.$$

It suffices to show that the second sum converges to 0 as $\delta \to 0$. Note that we can bound it by

$$\frac{1}{\delta} \sum_{i:S_i \subset N_{2\delta}(\beta)} \mathrm{diam}(Q_i) \, \mathrm{diam}(S_i) + \sum_{i \in D_\delta(\beta)} \mathrm{diam}(Q_i),$$

where $D_\delta(\beta)$ is the family of indices $i \in \mathbb{N}$ such that $\mathrm{diam}(S_i) \geq \delta$, $S_i \cap N_\delta(\beta) \neq \emptyset$, and $\overline{S}_i \cap \beta = \emptyset$. We are exactly in the setting of Lemma 3.6.9, so the conclusion follows. $\qquad \square$

It remains to prove the very first Lemma 3.9.6. We assume that we have a path β connecting two continua $E, F \subset [0, 1] \times [0, D]$, and its perturbations $\beta_s \subset \psi^{-1}(s) \cap ((0, 1) \times (0, D))$, for a.e. $s \in (0, \delta)$. We split the proof into two parts, stated in the following two lemmas.

Lemma 3.9.9 *For a.e. $s \in (0, \delta)$ the intersection $\psi^{-1}(s) \cap \partial S_i$ contains finitely many points, for all $i \in \mathbb{N} \cup \{0\}$.*

Lemma 3.9.10 *For a.e. $s \in (0, \delta)$ we have*

$$\mathcal{H}^1(g(\psi^{-1}(s) \cap \mathcal{R}^\circ)) = \mathcal{H}^1(g(\psi^{-1}(s)) \cap S^\circ) = 0.$$

Remark 3.9.11 One easily recognizes the connection to the proof of Lemma 3.5.11, where we first proved that for a.e. t the level set $u^{-1}(t)$ intersects the peripheral circles ∂Q_i in finitely many points (in fact in at most two points), and then treated separately the set $u^{-1}(t) \cap S^\circ$. In fact, Lemma 3.5.11 is a particular case of what we are about to show.

Proof of Lemma 3.9.9 The proof is based on the fact that $A := \partial S_i$ has finite Hausdorff 1-measure for all $i \in \mathbb{N} \cup \{0\}$. Let $J \subset (0, \delta)$ be the set of s for which $\psi^{-1}(s) \cap A$ contains at least N points, where $N \in \mathbb{N}$ is fixed. We will show that for the outer 1-measure $m_1^*(J)$ we have

$$m_1^*(J) \leq \frac{1}{N} \mathcal{H}^1(A). \tag{3.53}$$

Hence, the set of $s \in (0, \delta)$ for which $\psi^{-1}(s) \cap A$ contains infinitely many points has outer measure that is also bounded by $\frac{1}{N}\mathcal{H}^1(A)$, for all $N \in \mathbb{N}$. If we let $N \to \infty$ the conclusion will follow.

For $n \in \mathbb{N}$ define $J_n \subset J$ to be the set of $s \in (0, \delta)$ for which the intersection $\psi^{-1}(s) \cap A$ contains N points whose mutual distance is at least $1/n$; recall that N is fixed. It is easy to see that $J = \bigcup_{n \in \mathbb{N}} J_n$ and $J_n \subset J_{n+1}$, thus $m_1^*(J) = \lim_{n \to \infty} m_1^*(J_n)$; see [5, Prop. 1.5.12, p. 23]. It suffices to show (3.53) for J_n in the place of J.

Recall that $\psi(x) = \mathrm{dist}(x, \beta)$ for the given path β. For fixed $n \in \mathbb{N}$ we cover A by finitely many open sets $\{U_i\}_i$ with $\mathrm{diam}(U_i) < 1/n$. If $s \in J_n$ then $\psi^{-1}(s)$ meets at least N distinct sets U_i. For all t near s with $t \leq s$, the set $\psi^{-1}(t)$ also intersects at least N distinct sets U_i. Indeed, note that for each point $x \in \psi^{-1}(s)$ and for every small ball $B(x, r)$ there exists $t_0 < s$ such that $\psi^{-1}(t) \cap B(x, r) \neq \emptyset$ for all $t_0 \leq t \leq s$. To see the latter, let $[x, y]$ denote the line segment from x to its closest point $y \in \beta$, so $|x - y| = s$. Arbitrarily close to x we can find points $z \in [x, y]$, with $z \in \psi^{-1}(t_0)$ for some $t_0 < s$. Then by the intermediate value theorem applied to ψ on the segment $[z, x]$ the claim follows.

For $s \in J_n$ define W_s to be a non-trivial closed interval of the form $[t_0, s] \subset (0, \delta)$, such that $\psi^{-1}(t)$ intersects at least N distinct sets U_i for all $t \in W_s$. Also, let $W = \bigcup_{s \in J_n} W_s \supset J_n$, and $V_i = \psi(U_i)$. Then

$$\sum_i \chi_{V_i}(t) \geq N$$

for all $t \in W$. The set W is measurable as it is a union of non-trivial closed intervals; this is easy to see in one dimension, and a proof can be found in [3, Theorem 1.1]. Hence, we have

$$Nm_1^*(J_n) \leq Nm_1(W) = N \int \chi_W(t)\, dt \leq \int \sum_i \chi_{V_i}(t)\, dt$$

$$= \sum_i m_1(V_i) \leq \sum_i \mathrm{diam}(\psi(U_i)) \leq \sum_i \mathrm{diam}(U_i)$$

since ψ is 1-Lipschitz. The cover $\{U_i\}$ of A was arbitrary, so taking the infimum over all covers completes the proof. □

An ingredient for the proof of Lemma 3.9.10 is the following monotonicity property of f; cf. Lemma 3.4.7.

Lemma 3.9.12 *Let $x \in S$, $r > 0$, and $c > 1$ such that $B(x, r) \subset B(x, cr) \subset \Omega$. We have*

$$\mathrm{diam}(f(B(x, r) \cap S)) \leq \mathrm{diam}(f(B(x, sr) \cap S)) \leq \sum_{i:\overline{Q}_i \cap \partial B(x, sr) \neq \emptyset} \mathrm{diam}(S_i)$$

$$(3.54)$$

for a.e. $s \in [1, c]$.

Proof The map f is a homeomorphism so it has the *monotonicity* property that for any set $U \subset\subset \mathbb{C}$ we have

$$\sup_{z,w \in U} |f(z) - f(w)| = \sup_{z,w \in \partial U} |f(z) - f(w)|.$$

Thus, for each $s \in (1, c)$ we have

$$\text{diam}(f(B(x, r) \cap S)) \leq \text{diam}(f(B(x, sr) \cap S))$$

$$\leq \text{diam}(f(B(x, sr))) = |f(z) - f(w)|$$

for some points $z, w \in \partial B(x, sr)$. Suppose first that $z, w \in S$. Since for a.e. $s \in [1, c]$ the path $\partial B(x, sr)$ is non-exceptional for 2-modulus (this follows, for example, from a modification of Lemma 3.3.4), by the upper gradient inequality (3.48) we obtain

$$|f(z) - f(w)| \leq \sum_{i: \overline{Q}_i \cap \partial B(x, sr) \neq \emptyset} \text{diam}(S_i).$$

If z lies in a peripheral disk $\overline{Q}_{i_z}$ with $\overline{Q}_{i_z} \cap \partial B(x, sr) \neq \emptyset$, then we let z' be the last exit point of the path $\partial B(x, sr)$ from $\overline{Q}_{i_z}$ as it travels from z to w. Similarly, we consider a point w', in the case w lies in a peripheral disk $\overline{Q}_{i_w}$. Observe now that $|f(z) - f(z')| \leq \text{diam}(S_{i_z})$, $|f(w) - f(w')| \leq \text{diam}(S_{i_w})$, and the open subpath of $\partial B(x, sr)$ from z' to w' does not intersect $\overline{Q}_{i_z}, \overline{Q}_{i_w}$. The upper gradient inequality applied to this subpath yields the result. $\qquad\square$

Proof of Lemma 3.9.10 The proof is very similar to the proof of Lemma 3.5.11 so we omit most of the details.

For a fixed $\varepsilon > 0$ we consider the set $E_\varepsilon = \{i \in \mathbb{N} : \text{diam}(Q_i) > \varepsilon\}$. We cover $\Omega \setminus \bigcup_{i \in E_\varepsilon} \overline{Q}_i$ by balls B_j of radius $r_j < \varepsilon$ such that $2B_j \subset \Omega \setminus \bigcup_{i \in E_\varepsilon} \overline{Q}_i$, and such that $\frac{1}{5}B_j$ are disjoint.

Let J be the family of indices j such that for each $s \in [1, 2]$ we have

$$\text{diam}(f(sB_j \cap S)) \geq k \, \text{diam}(S_i)$$

for all peripheral disks Q_i with $\text{diam}(Q_i) > 8r_j$ that intersect $\partial(sB_j)$. The constant $k \geq 1$ can be chosen exactly as in the proof of Lemma 3.5.11. Using (3.54) for B_j and integrating over $s \in [1, 2]$ one obtains

$$r_j \, \text{diam}(f(B_j \cap S)) \leq C \sum_{i: Q_i \subset 11B_j} \text{diam}(S_i) \, \text{diam}(Q_i) \tag{3.55}$$

for all $j \in J$, where $C > 0$ is a uniform constant depending only on the data of the carpet S (cf. (3.16)). For each $j \in J$ consider the smallest interval I_j that contains $\psi(f(B_j \cap S))$ and define $g_\varepsilon(t) = \sum_{j \in J} 2r_j \chi_{I_j}(t)$, $t \in (0, \delta)$.

For $j \notin J$ there exists $s = s_j \in [1, 2]$ and there exists a peripheral disk Q_i that intersects $\partial(s B_j)$ with $\operatorname{diam}(Q_i) > 8r_j$, but $\operatorname{diam}(f(s B_j \cap S)) < k \operatorname{diam}(S_i)$. Let $\{Q_i\}_{i \in I}$ denote the family of such peripheral disks. Some Q_i, $i \in I$, might intersect multiple balls B_j, $j \notin J$. We define

$$\widetilde{Q}_i = \overline{Q}_i \cup \bigcup \left\{ s_j B_j : Q_i \cap \partial(s_j B_j) \neq \emptyset, \operatorname{diam}(Q_i) > 8r_j, \right.$$

$$\left. \text{and } \operatorname{diam}(f(s_j B_j \cap S)) < k \operatorname*{osc}_{Q_i}(u) \right\},$$

and note that

$$\operatorname{diam}(f(\widetilde{Q}_i \cap S)) \leq C \operatorname{diam}(S_i) \tag{3.56}$$

for all $i \in I$, where $C > 0$ depends only on the data. Furthermore, $\operatorname{diam}(\widetilde{Q}_i) < 2 \operatorname{diam}(Q_i)$ since $\operatorname{diam}(Q_i) > 8r_j$ whenever $s_j B_j \subset \widetilde{Q}_i$. Now, let I_i be the smallest interval containing $\psi(f(\widetilde{Q}_i \cap S))$ and define $b_\varepsilon(t) = \sum_{i \in I} 2 \operatorname{diam}(Q_i) \chi_{I_i}(t)$, $t \in (0, \delta)$.

Observe that for each $t \in (0, \delta)$ the set $g(\psi^{-1}(s) \cap \mathcal{R}^\circ)$ is covered by the balls B_j, $j \in J$, and the sets $\widetilde{Q}_i$, $i \in I$. Since $r_j < \varepsilon$ for $j \in J$ and $\operatorname{diam}(Q_i) < \varepsilon$ for $i \in I$, we have

$$\mathcal{H}_\varepsilon^1(g(\psi^{-1}(s) \cap \mathcal{R}^\circ)) \leq g_\varepsilon(s) + b_\varepsilon(s).$$

It suffices to show that $g_\varepsilon(s) \to 0$ and $b_\varepsilon(s) \to 0$ for a.e. $s \in (0, \delta)$, along a sequence of $\varepsilon \to 0$.

The function ψ is 1-Lipschitz, so we have

$$\operatorname{diam}(I_j) = \operatorname{diam}(\psi(f(B_j \cap S))) \leq \operatorname{diam}(f(B_j \cap S)) \qquad \text{and}$$

$$\operatorname{diam}(I_i) = \operatorname{diam}(\psi(f(\widetilde{Q}_i \cap S))) \leq \operatorname{diam}(f(\widetilde{Q}_i \cap S))$$

for all $j \in J$ and $i \in I$. These can be estimated above by (3.55) and (3.56), respectively. The proof continues exactly as in Lemma 3.5.11 by estimating $\int_0^\delta b_\varepsilon(s) \, ds$ and $\int_0^\delta g_\varepsilon(s) \, ds$ and showing that they converge to 0 as $\varepsilon \to 0$. $\qquad \square$

3.10 Carpet Modulus Estimates

In this section we state some modulus estimates, which were proved in [7, Sect. 8]. The statements there involve some bounds for the *transboundary modulus*, which is a notion of modulus that combines the classical conformal modulus and the carpet modulus that we are employing in this chapter. Thus, the proofs can be applied with minor changes and we restrict ourselves to mentioning the results.

We first recall some definitions from the Introduction of the current chapter (Sect. 3.1). Consider a Sierpiński carpet $S \subset \overline{\Omega}$ with its peripheral circles $\{\partial Q_i\}_{i \in \mathbb{N} \cup \{0\}}$, where $\partial Q_0 = \partial \Omega$. We say that the peripheral circles of the carpet S are K_2-quasicircles for some constant $K_2 > 0$ if ∂Q_i is a K_2-quasicircle for all $i \in \mathbb{N} \cup \{0\}$. This is to say for any two points $x, y \in \partial Q_i$ there exists an arc $\gamma \subset \partial Q_i$ connecting x and y with $\mathrm{diam}(\gamma) \leq K_3 |x - y|$. Furthermore, the peripheral circles of the carpet S are K_3-relatively separated for a constant $K_3 > 0$ if

$$\Delta(\partial Q_i, \partial Q_j) := \frac{\mathrm{dist}(\partial Q_i, \partial Q_j)}{\min\{\mathrm{diam}(\partial Q_i), \mathrm{diam}(\partial Q_j)\}} \geq K_3$$

for all $i, j \in \mathbb{N} \cup \{0\}$ with $i \neq j$. Recall that if the peripheral circles of S are uniform quasicircles then the inner peripheral disks Q_i, $i \in \mathbb{N}$, are uniformly Ahlfors 2-regular and uniformly quasiround.

In the following two propositions the common assumption is that we have a Sierpiński carpet S contained in a Jordan region $\overline{\Omega}$ such that $\partial Q_0 = \partial \Omega \subset S$ is the outer peripheral circle and $\{Q_i\}_{i \in \mathbb{N}}$ are the inner peripheral disks.

Proposition 3.10.1 ([7, Prop. 8.1]) *Assume that the peripheral circles $\{\partial Q_i\}_{i \in \mathbb{N} \cup \{0\}}$ of S are K_2-quasicircles and they are K_3-relatively separated, and fix an integer $N \in \mathbb{N}$. Then there exists a non-increasing function $\phi: (0, \infty) \to (0, \infty)$ that can be chosen only depending on K_2, K_3, and N with the following property: if E and F are arbitrary disjoint continua in S, and $I_0 \subset \mathbb{N} \cup \{0\}$ is a finite index set with $\#I_0 = N$, then the family of curves Γ in $\mathbb{C}$ joining the continua E and F, but avoiding the finitely many peripheral disks $\overline{Q}_i$, $i \in I_0$, has carpet modulus satisfying*

$$\mathrm{mod}(\Gamma) \geq \phi(\Delta(E, F)).$$

The same conclusion is true if we use instead the weak carpet modulus $\overline{\mathrm{mod}}(\Gamma)$.

Proposition 3.10.2 ([7, Prop. 8.4]) *Assume that the peripheral circles $\{\partial Q_i\}_{i \in \mathbb{N} \cup \{0\}}$ of S are K_2-quasicircles and they are K_3-relatively separated. Then there exists a non-increasing function $\phi: (0, \infty) \to (0, \infty)$ that can be chosen only depending on K_2 and K_3 with the following property: if E and F are disjoint continua in S, then the family of curves Γ in $\mathbb{C}$ joining the continua E and F has carpet modulus satisfying*

$$\mathrm{mod}(\Gamma) \leq \phi(\Delta(E, F)).$$

The same conclusion is true if we use instead the weak carpet modulus $\overline{\mathrm{mod}}(\Gamma)$.

A square Sierpiński carpet is by definition a Sierpiński carpet $\mathcal{R}$ whose inner peripheral disks $\{S_i\}_{i\in\mathbb{N}}$ are squares, and the outer peripheral circle ∂S_0 is a rectangle, where S_0 is the unbounded component of $\mathbb{C}\setminus\mathcal{R}$.

Proposition 3.10.3 ([7, Prop. 8.7]) *Let $\mathcal{R}\subset[0,1]\times[0,D]$ be a square Sierpiński carpet with inner peripheral squares $\{S_i\}_{i\in\mathbb{N}}$ and outer peripheral rectangle $\partial S_0 :=$ $\partial([0,1]\times[0,D])$. There exists a number $N = N(D) \in \mathbb{N}$ and a non-increasing function $\psi\colon (0,\infty) \to (0,\infty)$ with*

$$\lim_{t\to\infty} \psi(t) = 0$$

that can be chosen only depending on D and satisfies the following: if E and F are arbitrary continua in $\mathcal{R}$ with $\Delta(E,F) \geq 12$, then there exists a set $I_0 \subset \mathbb{N}\cup\{0\}$ with $\#I_0 \leq N$ such that the family of curves Γ in $\mathbb{C}$ joining the continua E and F, but avoiding the finitely many peripheral disks $\overline{S}_i$, $i \in I_0$, has carpet modulus satisfying

$$\mathrm{mod}(\Gamma) \leq \psi(\Delta(E,F)).$$

Moreover, if $D \in [D_1, D_2] \subset (0,\infty)$, the number N and the function ψ can be chosen to depend only on D_1, D_2.

As a last remark, (weak) carpet modulus of a path family Γ is always considered with respect to a given carpet, although this is not explicitly manifested in the notation $\mathrm{mod}(\Gamma)$. It will be clear from the context what the reference carpet each time is, when we use these estimates in the next section.

3.11   Quasisymmetric Uniformization

In this section we prove Theorem 3.1.2, which we restate for the convenience of the reader.

Theorem 3.11.1 *Let S be a Sierpiński carpet of area zero with peripheral circles $\{\partial Q_i\}_{i\in\mathbb{N}\cup\{0\}}$ that are K_2-quasicircles and K_3-relatively separated. Then there exists an η-quasisymmetric map f from S onto a square Sierpiński carpet $\mathcal{R}$ such that the distortion function η depends only on K_2 and K_3.*

Before proceeding to the proof we include some lemmas.

Lemma 3.11.2 *Let $a, b > 0$, and (X, d_X) and (Y, d_Y) be metric spaces. Suppose that $x_1, x_2, x_3 \in X$ and $y_1, y_2, y_3 \in Y$ are points such that*

$$d_X(x_i, x_j) \geq a \quad and \quad d_Y(y_i, y_j) \geq b \quad for \quad i, j = 1, 2, 3, \ i \neq j.$$

Then for all $x \in X$ and $y \in Y$ there exists an index $l \in \{1, 2, 3\}$ such that $d_X(x, x_l) \geq a/2$ and $d_Y(y, y_l) \geq b/2$.

Proof At most one of the points x_i can lie in the ball $B(x, a/2)$, so there are at least two points, say x_1, x_2 that have distance at least $a/2$ to x. At most one of the points y_1, y_2 can lie in $B(y, b/2)$, so one of them, say y_1, has to lie outside the ball $B(y, b/2)$. Then the desired statement holds for $l = 1$. $\square$

Lemma 3.11.3 *Let* $\mathcal{R} \subset [0, 1] \times [0, D]$ *be a square Sierpiński carpet such that* $\partial([0, 1] \times [0, D]) \subset \mathcal{R}$ *is the outer peripheral circle. Then there exists a constant* $C(D) > 0$ *such that the following two conditions are satisfied:*

1. *For all* $x, y \in \mathcal{R}$ *there exists a path* $\gamma \subset \mathcal{R}$ *connecting* x *and* y *with* $\mathrm{diam}(\gamma) \leq 2|x - y|$.
2. *If* $a \in \mathcal{R}$, $0 < r \leq C(D)$, *and* $x, y \in \mathcal{R} \setminus B(a, r)$, *then there exists a continuum* E *connecting* x *and* y *with* $E \subset \mathcal{R} \setminus B(a, r/2)$.

In fact, one can take $C(D) = \min\{1, D\}$.

Proof Let S_i, $i \in \mathbb{N}$, denote the inner (open) peripheral squares of $\mathcal{R}$, and $\partial S_0 := \partial([0, 1] \times [0, D])$. For the first statement note that for any two points x, y lying on a square ∂S_i there exists an arc $\gamma \subset \partial S_i$ connecting x, y with length at most $2|x - y|$. Now, if $x, y \in \mathcal{R}$ are arbitrary, we connect them with a line segment $[x, y]$, and then replace each of the segments $[x_i, y_i] := [x, y] \cap \overline{S}_i$ with an arc $\gamma_i \subset \partial S_i$ that connects x_i, y_i and has length at most $2|x_i - y_i|$. The resulting path $\gamma \subset \mathcal{R}$ connects x, y and has length at most $2|x - y|$.

For the second claim, let $C(D) = \min\{1, D\}$, so a ball $B(a, r/2)$ with $r \leq C(D)$ cannot intersect two opposite sides of the rectangle ∂S_0. If $x \in \mathcal{R} \setminus B(a, r)$, then we connect x to ∂S_0 with a line segment γ_x parallel to one of the coordinate axes that does not intersect $B(a, r)$. We replace each of the arcs $\gamma_x \cap S_i$, $i \in \mathbb{N}$, with an arc in ∂S_i that has the same endpoints and does not intersect $B(a, r)$. To see the existence of such an arc, note that if both of the arcs of ∂S_i with the same endpoints as $\gamma_x \cap S_i$ intersected $B(a, r)$, then $B(a, r)$ would also intersect $\gamma_x \cap S_i$, by convexity. This is a contradiction.

We still call the resulting path γ_x. We do the same for a point $y \in \mathcal{R} \setminus B(a, r)$ and obtain a path γ_y. Then one has to concatenate γ_x and γ_y with a path in ∂S_0 that does not intersect $B(a, r/2)$. If $\partial S_0 \setminus B(a, r/2)$ has only one component then this can be clearly done. The other case is that $\partial S_0 \setminus B(a, r/2)$ has two components E and F, and thus $B(a, r/2)$ intersects two neighboring sides of ∂S_0. The distance between a and these two sides is at most $r/2$, so $B(a, r)$ contains one of the components E, F, say it contains F. This now implies that the endpoints $\gamma_x \cap \partial S_0$, $\gamma_y \cap \partial S_0$ have to lie on E and can therefore be connected with a subarc of E. $\square$

Now we proceed to the proof of the main result. The candidate for the quasisymmetric map $f \colon S \to \mathcal{R}$ is the map that we constructed in the previous sections. The principle that we will use is that a "quasiconformal" map (in our case a map that preserves modulus) between a *Loewner space* and a space that is *linearly locally connected* (this is essentially implied by Lemma 3.11.3) is quasisymmetric; see [21, Chap. 11] for background. Certain complications arise since we do not know in advance that the peripheral squares of $\mathcal{R}$ are uniformly relatively separated, and we bypass this by employing Proposition 3.10.3.

Proof of Theorem 3.11.1 We apply the considerations from Sect. 3.4 to 3.9. So we obtain a homeomorphism f from S onto a square carpet $\mathcal{R} \subset [0, 1] \times [0, D]$, with outer peripheral disk $S_0 = \mathbb{C} \setminus [0, 1] \times [0, D]$. This homeomorphism maps $\partial \Omega$ to ∂S_0 and has the regularity as in Sect. 3.9: it satisfies the conclusions of Propositions 3.9.4 and 3.9.5; see also the formulation of Theorem 3.1.1.

We split the proof into two parts. The first part is to show that if we choose the sides $\Theta_1, \Theta_3 \subset \partial \Omega$ suitably, then the height D of the rectangle $[0, 1] \times [0, D]$ that contains the square carpet $\mathcal{R}$ is bounded above and below (away from 0), depending only on K_2 and K_3. Recall from Sect. 3.4.2 that the choice of Θ_1 and Θ_3 specifies uniquely the function u, and thus it specifies the uniformizing function $f = (u, v)$ and the height D of the rectangle that contains the carpet $\mathcal{R} = f(S)$. The second step is to prove that the map $f : S \to \mathcal{R}$ that satisfies the conclusions of Propositions 3.9.4 and 3.9.5 is a quasisymmetry.

For the first step, note that $\partial Q_0 = \partial \Omega$ is a K_2-quasicircle, so it is the quasisymmetric image of the unit circle. It follows that we can choose two disjoint arcs $\Theta_1, \Theta_3 \subset \partial \Omega$ with endpoints $a_i, i = 1, \ldots, 4$, such that

$$\min_{\substack{i \neq j \\ i,j=1,\ldots,4}} |a_i - a_j| \geq C_0 \operatorname{diam}(\partial \Omega) = C_0 \operatorname{diam}(S), \tag{3.57}$$

where $C_0 > 0$ is a constant depending only on K_2. Using this, and again that ∂Q_0 is a quasicircle, one can see that

$$\frac{1}{C'} \leq \Delta(\Theta_1, \Theta_3) \leq C'$$

for some constant $C' > 0$ depending only on K_2. Hence, if Γ denotes the family of paths in Ω that connect Θ_1 and Θ_3 (this family avoids $\overline{Q_0}$), by Proposition 3.10.1 and Proposition 3.10.2 we have

$$\operatorname{mod}(\Gamma) \leq C'' \quad \text{and} \quad \overline{\operatorname{mod}}(\Gamma) \geq C''' \tag{3.58}$$

for constants $C'', C''' > 0$ depending only on K_2 and K_3. The family $f(\Gamma)$ is the path family in $(0, 1) \times (0, D)$ that connects $\{0\} \times [0, D]$ to $\{1\} \times [0, D]$. Combining (3.58) with Propositions 3.9.5 and 3.9.4 we obtain

$$\overline{\operatorname{mod}}(f(\Gamma)) \leq C'' \quad \text{and} \quad \operatorname{mod}(f(\Gamma)) \geq C''',$$

where here (weak) carpet modulus is with respect to the carpet $\mathcal{R}$. Finally, we show that both of the above moduli are equal to D. Consider the discrete weight $\lambda(S_i) = \ell(S_i), i \in \mathbb{N}$, where $\ell(S_i)$ is the side length of the square S_i, and $\lambda(S_0) := 0$. It is immediate that $\{\lambda(S_i)\}_{i \in \mathbb{N} \cup \{0\}}$ is admissible for $f(\Gamma)$ with respect to both notions of modulus, since for any path $\gamma \in \Gamma$ with $\mathcal{H}^1(\gamma \cap \mathcal{R}) = 0$ we have

$$\sum_{i:Q_i \cap \gamma \neq \emptyset} \lambda(Q_i) \geq 1.$$

Thus, $\overline{\mathrm{mod}}(f(\Gamma))$ and $\mathrm{mod}(f(\Gamma))$ are both bounded above by

$$\sum_{i \in \mathbb{N}} \ell(S_i)^2 = \mathcal{H}^2([0, 1] \times [0, D]) = D.$$

Conversely, if $\lambda(S_i)$ is an arbitrary weight that is admissible for $\overline{\mathrm{mod}}(f(\Gamma))$, then we may assume that $\lambda(S_0) = 0$, since the path family $f(\Gamma)$ does not hit $\overline{S}_0$. Moreover, for the paths $\gamma_t(r) = (r, t)$, $r \in (0, 1)$, and for a.e. $t \in [0, 1]$ we have

$$1 \leq \sum_{i:\overline{S}_i \cap \gamma_t \neq \emptyset} \lambda(S_i).$$

This is because the paths $\gamma_t(r)$ are non-exceptional for a.e. $t \in [0, 1]$; see Lemma 3.3.4 and its proof in Lemma 2.4.3.

Integrating over $t \in [0, 1]$ and applying Fubini's theorem yields

$$1 \leq \sum_{i \in \mathbb{N}} \lambda(S_i) \int_0^1 \chi_{\overline{S}_i \cap \gamma_t}\, dt \leq \sum_{i \in \mathbb{N}} \lambda(S_i)\ell(S_i) \leq \left(\sum_{i \in \mathbb{N}} \lambda(S_i)^2\right)^{1/2} \left(\sum_{i \in \mathbb{N}} \ell(S_i)^2\right)^{1/2}$$

$$= \left(\sum_{i \in \mathbb{N}} \lambda(S_i)^2\right)^{1/2} \mathcal{H}^2([0, 1] \times [0, D])^{1/2}.$$

Hence $\sum_{i \in \mathbb{N}} \lambda(S_i)^2 \geq D$, which shows that $\overline{\mathrm{mod}}(f(\Gamma)) \geq D$. The same computation proves the claim for $\mathrm{mod}(f(\Gamma))$. Summarizing, under our choice of Θ_1 and Θ_3, the height D of the rectangle $(0, 1) \times (0, D)$ is bounded above and below, depending only on K_2 and K_3.

Now we move to the second step of the main proof. We remark that among the constants that we introduced in the first step only the constant C_0 is used again, and all other constants here are new constants. For simplicity we rescale S so that $\mathrm{diam}(S) = 1$. This does not affect the constants K_2, K_3, or the quasisymmetry distortion function. We show that the map $f: S \to \mathcal{R}$ is a *weak quasisymmetry*, i.e., there exists a constant $H > 0$ depending only on K_2, K_3 such that for any three points $x, y, z \in S$ with $|x - y| \leq |x - z|$ we have that the images $x' = f(x)$, $y' = f(y)$, $z' = f(z)$ satisfy

$$|x' - y'| \leq H|x' - z'|.$$

By a well-known criterion this implies that f is an η-quasisymmetry, where η depends only on K_2, K_3; see [21, Theorem 10.19].

We argue by contradiction, assuming that there exist points $x, y, z \in S$ with $|x - y| \leq |x - z|$, but $|x' - y'| > H|x' - z'|$ for some large $H > 0$. Then the points x, y, z are distinct. By Lemma 3.11.3(1), there exists a continuum $E' \subset \mathcal{R}$ with $\mathrm{diam}(E') \leq 2|x' - z'|$, connecting x', z'.

Recall that the distance between the endpoints a_i, $i = 1, \ldots, 4$, of Θ_1 and Θ_3 is at least

$$\delta := \min_{\substack{i \neq j \\ i,j=1,\ldots,4}} |a_i - a_j| \geq C_0$$

by (3.57). Their images a'_i, $i = 1, \ldots, 4$, are the vertices of the rectangle $\partial S_0 = \partial([0, 1] \times [0, D])$. So their mutual distance is bounded below by $\delta' = \min\{1, D\}$. By Lemma 3.11.2, there exists an index $i = 1, \ldots, 4$ such that for $u = a_i$ and $u' = a'_i$ we have

$$|u - y| \geq \delta/2 \quad \text{and} \quad |u' - x'| \geq \delta'/2.$$

Since $|x' - y'| \leq 2\max\{1, D\}$, it follows that

$$|u' - x'| \geq \frac{\delta'}{2} \geq \frac{1}{4}\frac{\min\{1, D\}}{\max\{1, D\}}|x' - y'| = \frac{1}{4}\min\{D, 1/D\}|x' - y'|$$
$$=: C_1(D)|x' - y'|.$$

Hence, $u' \notin B(x', r)$, where $r = C_1(D)|x' - y'|$. The fact that $C_1(D) < 1$ implies that we also have $y' \notin B(x', r)$. By Lemma 3.11.3(2), we can find a continuum $F' \subset \mathcal{R} \setminus B(x', r/2)$, connecting u' and y'. We have

$$\mathrm{dist}(E', F') \geq \frac{r}{2} - \mathrm{diam}(E') \geq \frac{C_1(D)}{2}|x' - y'| - 2|x' - z'|$$
$$\geq \left(H\frac{C_1(D)}{2} - 2\right)|x' - z'|$$
$$\geq \frac{HC_1(D)}{4}|x' - z'|$$

for $H \geq \frac{8}{C_1(D)}$. Also,

$$\min\{\mathrm{diam}(E'), \mathrm{diam}(F')\} \leq \mathrm{diam}(E') \leq 2|x' - z'|.$$

Therefore,

$$\Delta(E', F') = \frac{\text{dist}(E', F')}{\min\{\text{diam}(E'), \text{diam}(F')\}} \geq \frac{HC_1(D)}{8} \geq H \cdot C',$$

where the constant $C' > 0$ depends only on the lower and upper bounds of D, and thus only on K_2 and K_3, by the first step of the proof.

By making our initial choice of H sufficiently large, we may assume that $HC' \geq 12$ and apply Proposition 3.10.3. With N and ψ as in Proposition 3.10.3, there exists an index set $I_0 \subset \mathbb{N} \cup \{0\}$ with $\#I_0 \leq N$ such that the family Γ' of curves joining E' and F' in $\mathbb{C}$ but avoiding the peripheral disks $\overline{S}_i$, $i \in I_0$, satisfies

$$\text{mod}(\Gamma') \leq \psi(\Delta(E', F')) \leq \psi(HC'). \tag{3.59}$$

The number $N \in \mathbb{N}$ and the function ψ depend only on D and thus, only on K_2 and K_3.

Define $E = f^{-1}(E')$ and $F = f^{-1}(F')$. Then E and F are disjoint continua in S containing the sets $\{x, z\}$ and $\{y, u\}$, respectively. We have

$$\text{diam}(F) \geq |y - u| \geq \frac{\delta}{2} \geq \frac{\delta}{2} \, \text{diam}(E) \geq \frac{C_0}{2} \, \text{diam}(E),$$

since $\text{diam}(E) \leq \text{diam}(S) = 1$. Also,

$$\text{dist}(E, F) \leq |x - y| \leq |x - z| \leq \text{diam}(E)$$

$$\leq \max\left\{1, \frac{2}{C_0}\right\} \cdot \min\{\text{diam}(E), \text{diam}(F)\}.$$

Therefore,

$$\Delta(E, F) \leq \max\left\{1, \frac{2}{C_0}\right\} =: C'',$$

where the latter is a uniform constant. The path family $\Gamma = f^{-1}(\Gamma')$ connects the continua E and F in $\mathbb{C}$ but avoids the finitely many peripheral disks $\overline{Q}_i$, $i \in I_0$. Since $\#I_0$ is uniformly bounded, depending only on K_2 and K_3, by Proposition 3.10.1 we have

$$\overline{\text{mod}}(\Gamma) \geq \phi(\Delta(E, F)) \geq \phi(C''). \tag{3.60}$$

Combining (3.60) and (3.59) with the modulus inequality for f in Proposition 3.9.4 we obtain

$$\phi(C'') \leq \overline{\text{mod}}(\Gamma) \leq \text{mod}(f(\Gamma)) = \text{mod}(\Gamma') \leq \psi(HC').$$

Since $\lim_{t \to \infty} \psi(t) = 0$, if H is sufficiently large depending only on K_2 and K_3, we obtain a contradiction. $\square$

3.12   Equivalence of Square and Round Carpets

In this section we prove Proposition 3.1.6.

Proof of Proposition 3.1.6 We denote the peripheral circles of a square carpet $\mathcal{R}$ by ∂S_i, $i \in \mathbb{N} \cup \{0\}$, and the ones corresponding to the round carpet T by ∂C_i, $i \in \mathbb{N} \cup \{0\}$.

If the peripheral circles of a square carpet $\mathcal{R}$ are uniformly relatively separated, then by Bonk's result in Theorem 3.1.4, $\mathcal{R}$ is quasisymmetrically equivalent to a round carpet T. Conversely if the peripheral circles of a round carpet T are uniformly relatively separated, then our main result Theorem 3.1.2 implies that T is quasisymmetrically equivalent to a square carpet. We remark here that the property of uniform relative separation is a quasisymmetric invariant.

Now, assume that we are given a quasisymmetry F from a square carpet $\mathcal{R}$ onto a round carpet T, but the peripheral circles ∂S_i, $i \in \mathbb{N} \cup \{0\}$, of $\mathcal{R}$ are not uniformly relatively separated. This implies that there exists a sequence of pairs W_n, Z_n of rectangles (i.e., $W_n = \partial S_i$ and $Z_n = \partial S_j$ for some $i, j \in \mathbb{N} \cup \{0\}$) such that $\Delta(W_n, Z_n) \to 0$ as $n \to \infty$. Note that the quasisymmetry F maps each peripheral circle of $\mathcal{R}$ onto a peripheral circle of T.

Consider the largest parallel line segments $\alpha_n \subset W_n$ and $\beta_n \subset Z_n$ such that one of them is either a horizontal *or* vertical translation of the other. Let $d_n := \mathrm{dist}(W_n, Z_n) = \mathrm{dist}(\alpha_n, \beta_n)$, and define ℓ_n to be the length of α_n and β_n. If such segments do not exist, then we set $\alpha_n \in W_n$ and $\beta_n \in Z_n$ to be the nearest vertices of the two rectangles. In this case we set $\ell_n = 0$. We split into two cases.

Case 1 W_n and Z_n "tend" to share a large segment of a side. More precisely, we assume in this case that $\lim_{n \to \infty} d_n/\ell_n = 0$. In particular, $\ell_n \neq 0$ for all but finitely many n. By passing to a subsequence, we assume that this is the case for all $n \in \mathbb{N}$.

Consider a K-quasiconformal extension of F to $\widehat{\mathbb{C}}$, as in [7, Sect. 5]. Let g_n be a Euclidean similarity that maps $U := [0, 1] \times \{0\}$ onto β_n. Also, let h_n be a Möbius transformation that maps the disk bounded by the circle $F(Z_n)$ onto $\overline{\mathbb{D}}$ such that $f_n := h_n \circ F \circ g_n$ satisfies $f_n(0) = 1$, $f_n(1) = -1$, and $f_n(1/2) = i$ or $f_n(1/2) = -i$ so that f is orientation-preserving. Note that $f_n : \widehat{\mathbb{C}} \to \widehat{\mathbb{C}}$ is K-quasiconformal. By passing to a subsequence, we assume that f_n converges uniformly in the spherical metric of $\widehat{\mathbb{C}}$ to a K-quasiconformal map $f : \widehat{\mathbb{C}} \to \widehat{\mathbb{C}}$; see [25, Theorem 5.1, p. 73] for compactness properties of quasiconformal maps.

Since g_n is a scaling by ℓ_n, it follows that the Euclidean distance between $U = g_n^{-1}(\beta_n)$ and $V_n := g_n^{-1}(\alpha_n)$ is d_n/ℓ_n, and in fact V_n is a horizontal line segment of length 1. In what follows, we will use Hausdorff convergence with respect to the spherical metric of $\widehat{\mathbb{C}}$, and the fact that the Hausdorff convergence is compatible with the uniform convergence of f_n; for instance, V_n converges to U in the Hausdorff sense, and thus $f_n(V_n)$ converges to $f(U)$ in the Hausdorff sense.

The idea now is that the rectangles $g_n^{-1}(Z_n)$, $g_n^{-1}(W_n)$ converge to two sets that share a line segment, but the images of $g_n^{-1}(Z_n)$, $g_n^{-1}(W_n)$ under f_n converge to two tangent circles that share only one point. This contradicts the fact that f is a homeomorphism.

To prove the above claims, observe first that any Hausdorff limit of the sets $g_n^{-1}(Z_n) \cup g_n^{-1}(W_n) \subset \widehat{\mathbb{C}}$ is not a circle in $\widehat{\mathbb{C}}$. On the other hand, $f_n(V_n)$ is an arc of a circle, so its Hausdorff limit will be an arc of a circle $C \subset \widehat{\mathbb{C}}$. The circle C is distinct from $\partial \mathbb{D}$ and they bound disjoint regions; this is justified because f is a homeomorphism and the Hausdorff limit of $g_n^{-1}(Z_n) \cup g_n^{-1}(W_n)$ is not just a single circle. Thus, $C \cap \partial \mathbb{D}$ can contain at most one point. On the other hand, $f(U) = \lim_{n \to \infty} f_n(U) = \lim_{n \to \infty} f_n(V_n) \subset C \cap \partial \mathbb{D}$, which a contradiction, since f is a homeomorphism and cannot map U to a point.

Case 2 W_n and Z_n "tend" to share a corner. More precisely, we assume that there exists a constant $c > 0$ such that $d_n/\ell_n \geq c$ for infinitely many n. We allow the possibility $d_n/\ell_n = \infty$, which occurs whenever α_n and β_n are vertices and $\ell_n = 0$. The assumption that $\Delta(W_n, Z_n) \to 0$ implies that in this case neither W_n nor Z_n can be the outer peripheral rectangle ∂S_0, so they are both squares, for sufficiently large n. By passing to a subsequence we assume that the above hold for all $n \in \mathbb{N}$. Also, assume that Z_n is smaller than W_n, so if m_n denotes the side length of Z_n, then $d_n/m_n \to 0$ as $n \to \infty$ by our assumption on the separation of Z_n and W_n. Note that we also have $\ell_n/m_n \to 0$ as $n \to \infty$.

Again, we consider a quasiconformal extension $F \colon \widehat{\mathbb{C}} \to \widehat{\mathbb{C}}$. We precompose F with a Euclidean similarity g_n that maps the unit square $U = \partial([0, 1] \times [0, 1])$ onto the square Z_n, and postcompose with a Möbius transformation h_n that maps the disk bounded by the circle $F(Z_n)$ onto $\overline{\mathbb{D}}$, with suitable normalizations. After passing to a subsequence we may assume that $f_n := h_n \circ F \circ g_n$ converges uniformly in the spherical metric to a K-quasiconformal map $f \colon \widehat{\mathbb{C}} \to \widehat{\mathbb{C}}$.

Note that the Euclidean distance between $U = g_n^{-1}(Z_n)$ and $V_n := g_n^{-1}(W_n)$ is d_n/m_n, the square V_n is larger than U, and the segments $g^{-1}(\alpha_n)$ and $g^{-1}(\beta_n)$ have length ℓ_n/m_n, which converges to 0. After passing to a subsequence, it follows that a Hausdorff limit (in $\widehat{\mathbb{C}}$) of V_n contains two perpendicular segments of Euclidean length at least 1 that meet at a corner x of U, but otherwise disjoint from U. Also, observe that all the arcs that lie in the union of these two segments with ∂U are quasiarcs.

On the other hand, the image $f_n(V_n)$ converges to a circle $C \subset \widehat{\mathbb{C}}$ that meets $\partial \mathbb{D}$ at one point $x' = f(x)$. It is easy to see that there exist arbitrarily small arcs in $C \cup \partial \mathbb{D}$ passing through x' that are not quasiarcs. This leads to a contradiction, since the quasiconformal map $f \colon \widehat{\mathbb{C}} \to \widehat{\mathbb{C}}$ is a quasisymmetry and thus must map quasiarcs to quasiarcs.

As a final remark, if we are given a quasisymmetry F between a square carpet $\mathcal{R}$ and a round carpet T, but the peripheral circles ∂C_i, $i \in \mathbb{N} \cup \{0\}$, of T are not uniformly relatively separated, then the peripheral circles ∂S_i, $i \in \mathbb{N} \cup \{0\}$, of $\mathcal{R}$ are also not uniformly relatively separated, so we are reduced to the previous analysis. The proof is complete. $\square$

3.13   A Test Function

Here we include a lemma that is often used in variational arguments in Sect. 3.6. The assumptions here are that we have a carpet $S \subset \overline{\Omega}$ of area zero with outer peripheral circle $\partial Q_0 = \partial \Omega$ and inner peripheral disks $\{Q_i\}_{i \in \mathbb{N}}$ that are uniformly Ahlfors 2-regular and uniformly quasiround.

Lemma 3.13.1 *For each $x \in S$, $r > 0$, and $\varepsilon > 0$ there exists a function $\zeta \in \mathcal{W}^{1,2}(S)$, supported in $B(x, r) \cap S$, with $0 \leq \zeta \leq 1$ and $\zeta \equiv 1$ in some smaller ball $B(x, r') \cap S$, $r' < r$, such that:*

(a) *If $x \in S^\circ \cup \partial \Omega$ then*

$$D(\zeta) = \sum_{i \in \mathbb{N}} \underset{Q_i}{\mathrm{osc}}(\zeta)^2 < \varepsilon.$$

(b) *If $x \in \partial Q_{i_0}$ for some $i_0 \in \mathbb{N}$ then*

$$D(\zeta) - \underset{Q_{i_0}}{\mathrm{osc}}(\zeta)^2 = \sum_{i \in \mathbb{N} \setminus \{i_0\}} \underset{Q_i}{\mathrm{osc}}(\zeta)^2 < \varepsilon.$$

Proof The function ζ will be a discrete version of the logarithm. In fact we will construct a Lipschitz function ζ defined on all of $\mathbb{R}^2$, and thus it will lie in $\mathcal{W}^{1,2}(S)$; see comments after Definition 3.4.2 and also Example 2.5.19.

   We fix a large integer N which will correspond to the number of annuli around x that we will construct, and ζ will increase by $1/N$ on each annulus. We set $\zeta = 0$ outside $B(x, r)$ and define $R_1 = r$ and $r_1 = R_1/2$. In the annulus $A_1 = A(x; r_1, R_1)$ define ζ to be the radial function of constant slope $\frac{1}{N r_1}$, so that on the inner boundary of A_1 the function ζ has value $\frac{1}{N}$. Then consider $R_2 < r_1$ sufficiently small, and $r_2 = R_2/2$, so that no peripheral disk intersects both annuli A_1 and $A_2 := A(x; r_2, R_2)$, except possibly for Q_{i_0}, in the case $x \in \partial Q_{i_0}$; this is possible since the diameters of the peripheral disks converge to 0. In the transition annulus $A(x; R_2, r_1)$ we define ζ to be constant, equal to $\frac{1}{N}$, and on A_2 we let ζ be the radial function of slope $\frac{1}{N r_2}$. We continue constructing annuli $A_j = A(x; r_j, R_j)$, $j = 1, \dots, N$, and defining the function ζ in the same way. The last annulus will be $A_N := A(x; r_N, R_N)$ and the value of ζ will be 1 in the inner boundary of A_N. We extend ζ to be 1 in the ball $B(x, r_N)$.

   We now compute the Dirichlet energy $D(\zeta)$ of ζ. Assume that $x \in \partial Q_{i_0}$, $i_0 \in \mathbb{N}$, since otherwise the details are almost the same, but simpler. For $i \in \mathbb{N}$ and

$j \in \{1, \ldots, N\}$, let $d_j(Q_i) = \mathcal{H}^1(\{s \in [r_j, R_j] : \gamma_s \cap Q_i \neq \emptyset\})$, where γ_s is the circle of radius s around x. Since the peripheral disks Q_i, $i \in \mathbb{N}$, are Ahlfors 2-regular, there exists a constant $C > 0$ such that $d_j(Q_i)^2 \leq C\mathcal{H}^2(Q_i \cap A_j)$, for all $i \in \mathbb{N}$ and $j \in \{1, \ldots, N\}$; see Remark 3.5.10. Also, if $Q_i \cap A_j \neq \emptyset$, $i \neq i_0$, then $\mathrm{osc}_{Q_i}(\zeta) \leq d_j(Q_i)\frac{1}{Nr_j}$. By construction, each peripheral disk Q_i, $i \neq i_0$, can only intersect one annulus A_j, and if a peripheral disk Q_i does not intersect any annulus A_j, then ζ is constant on Q_i, so $\mathrm{osc}_{Q_i}(\zeta) = 0$. Combining these observations, we have

$$
\sum_{i \in \mathbb{N} \setminus \{i_0\}} \mathrm{osc}_{Q_i}(\zeta)^2 = \sum_{j=1}^{N} \sum_{\substack{i : Q_i \cap A_j \neq \emptyset \\ i \in \mathbb{N} \setminus \{i_0\}}} \mathrm{osc}_{Q_i}(\zeta)^2 \leq \frac{1}{N^2} \sum_{j=1}^{N} \frac{1}{r_j^2} \sum_{\substack{i : Q_i \cap A_j \neq \emptyset \\ i \in \mathbb{N} \setminus \{i_0\}}} d_j(Q_i)^2
$$

$$
\leq \frac{C}{N^2} \sum_{j=1}^{N} \frac{1}{r_j^2} \sum_{\substack{i : Q_i \cap A_j \neq \emptyset \\ i \in \mathbb{N} \setminus \{i_0\}}} \mathcal{H}^2(Q_i \cap A_j)
$$

$$
\leq \frac{C}{N^2} \sum_{j=1}^{N} \frac{1}{r_j^2} \mathcal{H}^2(A_j) = \frac{\pi C}{N^2} \sum_{j=1}^{N} \frac{4r_j^2 - r_j^2}{r_j^2}
$$

$$
= \frac{3\pi C}{N}.
$$

Making N sufficiently large we can achieve that $\frac{3\pi C}{N} < \varepsilon$, as desired. $\square$

References

1. L. Ambrosio, P. Tilli, *Topics on Analysis in Metric Spaces*. Oxford Lecture Series in Mathematics and its Applications, vol. 25 (Oxford University Press, Oxford, 2004)
2. K. Astala, T. Iwaniec, G. Martin, *Elliptic Partial Differential Equations and Quasiconformal Mappings in the Plane*. Princeton Mathematical Series, vol. 48 (Princeton University Press, Princeton, 2009)
3. M. Balcerzak, A. Kharazishvili, On uncountable unions and intersections of measurable sets. Georgian Math. J. **6**(3), 201–212 (1999)
4. Z.M. Balogh, I. Holopainen, J.T. Tyson, Singular solutions, homogeneous norms, and quasiconformal mappings in Carnot groups. Math. Ann. **324**(1), 159–186 (2002)
5. V.I. Bogachev, *Measure Theory*, vol. I (Springer, Berlin, 2007)
6. B. Bojarski, Remarks on Sobolev imbedding inequalities, in *Complex Analysis, Joensuu 1987*. Lecture Notes in Mathematics, vol. 1351 (Springer, Berlin, 1988), pp. 52–68
7. M. Bonk, Uniformization of Sierpiński carpets in the plane. Invent. Math. **186**(3), 559–665 (2011)
8. M. Bonk, B. Kleiner, Quasisymmetric parametrizations of two-dimensional metric spheres. Invent. Math. **150**(1), 127–183 (2002)
9. M. Bonk, S. Merenkov, Quasisymmetric rigidity of square Sierpiński carpets. Ann. Math. **177**(2), 591–643 (2013)
10. M. Brown, Sets of constant distance from a planar set. Michigan Math. J. **19**, 321–323 (1972)
11. R.B. Burckel, *An Introduction to Classical Complex Analysis, vol. 1*. Pure and Applied Mathematics, vol. 82 (Academic Press (Harcourt Brace Jovanovich, Publishers), New York; London, 1979)
12. J.W. Cannon, The combinatorial Riemann mapping theorem. Acta Math. **173**(2), 155–234 (1994)
13. J. Cheeger, Differentiability of Lipschitz functions on metric measure spaces. Geom. Funct. Anal. **9**(3), 428–517 (1999)
14. R. Courant, *Dirichlet's Principle, Conformal Mapping, and Minimal Surfaces* (Springer, New York-Heidelberg, 1977). Reprint of the 1950 original
15. R.J. Daverman, *Decompositions of Manifolds*. Pure and Applied Mathematics, vol. 124 (Academic Press, Orlando, 1986)
16. P. Duren, *Harmonic Mappings in the Plane*. Cambridge Tracts in Mathematics, vol. 156 (Cambridge University Press, Cambridge, 2004)
17. H. Federer, *Geometric Measure Theory*. Grundlehren der mathematischen Wissenschaften, vol. 153 (Springer, New York, 1969)

© The Editor(s) (if applicable) and The Author(s), under exclusive licence to Springer Nature Switzerland AG 2020
D. Ntalampekos, *Potential Theory on Sierpiński Carpets*, Lecture Notes in Mathematics 2268, https://doi.org/10.1007/978-3-030-50805-0

18. G.B. Folland, *Real Analysis. Modern Techniques and Their Applications*. Pure and Applied Mathematics (New York), 2nd edn. (Wiley, New York, 1999)
19. S. Granlund, Harnack's inequality in the borderline case. Ann. Acad. Sci. Fenn. Ser. A I Math. **5**(1), 159–163 (1980)
20. S. Granlund, N. Marola, On the problem of unique continuation for the p-Laplace equation. Nonlinear Anal. **101**, 89–97 (2014)
21. J. Heinonen, *Lectures on Analysis on Metric Spaces*. Universitext (Springer, New York, 2001)
22. J. Heinonen, P. Koskela, N. Shanmugalingam, J.T. Tyson, *Sobolev Spaces on Metric Measure Spaces. An Approach Based on Upper Gradients*. New Mathematical Monographs, vol. 27 (Cambridge University Press, Cambridge, 2015)
23. T. Iwaniec, L.V. Kovalev, J. Onninen, Hopf differentials and smoothing Sobolev homeomorphisms. Int. Math. Res. Not. **2012**(14), 3256–3277 (2012)
24. M. Kapovich, B. Kleiner, Hyperbolic groups with low-dimensional boundary. Ann. Sci. École Norm. Sup. **33**(5), 647–669 (2000)
25. O. Lehto, K. I. Virtanen, *Quasiconformal Mappings in the Plane*. Grundlehren der mathematischen Wissenschaften, vol. 126, 2nd edn. (Springer, New York, 1973)
26. A. Lytchak, S. Wenger, Canonical parameterizations of metric disks. Duke Math. J. **169**(4), 761–797 (2020)
27. S. Merenkov, Planar relative Schottky sets and quasisymmetric maps. Proc. Lond. Math. Soc. **104**(3), 455–485 (2012)
28. S. Merenkov, Local rigidity of Schottky maps. Proc. Am. Math. Soc. **142**(12), 4321–4332 (2014)
29. R.L. Moore, Concerning upper semi-continuous collections of continua. Trans. Am. Math. Soc. **27**(4), 416–428 (1925)
30. D. Ntalampekos, Monotone Sobolev functions in planar domains: level sets and smooth approximation (2019). arXiv:1911.06619
31. C. Pommerenke, *Boundary Behaviour of Conformal Maps*. Grundlehren der mathematischen Wissenschaften, vol. 299 (Springer, Berlin, 1992)
32. T. Rado, P.V. Reichelderfer, *Continuous Transformations in Analysis. With an Introduction to Algebraic Topology*. Grundlehren der mathematischen Wissenschaften, vol. 75 (Springer, Berlin, 1955)
33. K. Rajala, Uniformization of two-dimensional metric surfaces. Invent. Math. **207**(3), 1301–1375 (2017)
34. R. Remmert, *Classical Topics in Complex Function Theory*. Graduate Texts in Mathematics, vol. 172 (Springer, New York, 1998)
35. B. Rodin, D. Sullivan, The convergence of circle packings to the Riemann mapping. J. Differ. Geom. **26**(2), 349–360 (1987)
36. O. Schramm, Square tilings with prescribed combinatorics. Israel J. Math. **84**(1–2), 97–118 (1993)
37. O. Schramm, Transboundary extremal length. J. Anal. Math. **66**, 307–329 (1995)
38. N. Shanmugalingam, Newtonian spaces: an extension of Sobolev spaces to metric measure spaces. Rev. Mat. Iberoamericana **16**(2), 243–279 (2000)
39. N. Shanmugalingam, Harmonic functions on metric spaces. Ill. J. Math. **45**(3), 1021–1050 (2001)
40. W. Smith, D.A. Stegenga, Hölder domains and Poincaré domains. Trans. Am. Math. Soc. **319**(1), 67–100 (1990)
41. D. Sullivan, On the ergodic theory at infinity of an arbitrary discrete group of hyperbolic motions, in *Riemann Surfaces and Related Topics: Proceedings of the 1978 Stony Brook Conference (State University of New York, Stony Brook, 1978)*. Annals of Mathematics Studies, vol. 97 (Princeton University Press, Princeton, 1981), pp. 465–496
42. J.T. Tyson, Metric and geometric quasiconformality in Ahlfors regular Loewner spaces. Conform. Geom. Dyn. **5**, 21–73 (2001)
43. J. Väisälä, *Lectures on n-Dimensional Quasiconformal Mappings*. Lecture Notes in Mathematics, vol. 229 (Springer, Berlin, 1971)

44. G.T. Whyburn, Topological characterization of the Sierpiński curve. Fund. Math. **45**, 320–324 (1958)
45. R.L. Wilder, *Topology of Manifolds*. American Mathematical Society Colloquium Publications, vol. 32 (American Mathematical Society, Providence, 1963)
46. S. Willard, *General Topology* (Addison-Wesley, Reading, 1970)
47. K. Yosida, *Functional Analysis*. Grundlehren der mathematischen Wissenschaften, vol. 123, 6th edn. (Springer, Berlin, 1980)

Index

© The Editor(s) (if applicable) and The Author(s), under exclusive licence
to Springer Nature Switzerland AG 2020
D. Ntalampekos, *Potential Theory on Sierpiński Carpets*, Lecture Notes
in Mathematics 2268, https://doi.org/10.1007/978-3-030-50805-0

LECTURE NOTES IN MATHEMATICS

Editors in Chief: J.-M. Morel, B. Teissier;

Editorial Policy

1. Lecture Notes aim to report new developments in all areas of mathematics and their applications – quickly, informally and at a high level. Mathematical texts analysing new developments in modelling and numerical simulation are welcome.

   Manuscripts should be reasonably self-contained and rounded off. Thus they may, and often will, present not only results of the author but also related work by other people. They may be based on specialised lecture courses. Furthermore, the manuscripts should provide sufficient motivation, examples and applications. This clearly distinguishes Lecture Notes from journal articles or technical reports which normally are very concise. Articles intended for a journal but too long to be accepted by most journals, usually do not have this "lecture notes" character. For similar reasons it is unusual for doctoral theses to be accepted for the Lecture Notes series, though habilitation theses may be appropriate.

2. Besides monographs, multi-author manuscripts resulting from SUMMER SCHOOLS or similar INTENSIVE COURSES are welcome, provided their objective was held to present an active mathematical topic to an audience at the beginning or intermediate graduate level (a list of participants should be provided).

   The resulting manuscript should not be just a collection of course notes, but should require advance planning and coordination among the main lecturers. The subject matter should dictate the structure of the book. This structure should be motivated and explained in a scientific introduction, and the notation, references, index and formulation of results should be, if possible, unified by the editors. Each contribution should have an abstract and an introduction referring to the other contributions. In other words, more preparatory work must go into a multi-authored volume than simply assembling a disparate collection of papers, communicated at the event.

3. Manuscripts should be submitted either online at www.editorialmanager.com/lnm to Springer's mathematics editorial in Heidelberg, or electronically to one of the series editors. Authors should be aware that incomplete or insufficiently close-to-final manuscripts almost always result in longer refereeing times and nevertheless unclear referees' recommendations, making further refereeing of a final draft necessary. The strict minimum amount of material that will be considered should include a detailed outline describing the planned contents of each chapter, a bibliography and several sample chapters. Parallel submission of a manuscript to another publisher while under consideration for LNM is not acceptable and can lead to rejection.

4. In general, **monographs** will be sent out to at least 2 external referees for evaluation.

   A final decision to publish can be made only on the basis of the complete manuscript, however a refereeing process leading to a preliminary decision can be based on a pre-final or incomplete manuscript.

   Volume Editors of **multi-author works** are expected to arrange for the refereeing, to the usual scientific standards, of the individual contributions. If the resulting reports can be forwarded to the LNM Editorial Board, this is very helpful. If no reports are forwarded or if other questions remain unclear in respect of homogeneity etc, the series editors may wish to consult external referees for an overall evaluation of the volume.

5. Manuscripts should in general be submitted in English. Final manuscripts should contain at least 100 pages of mathematical text and should always include

   - a table of contents;
   - an informative introduction, with adequate motivation and perhaps some historical remarks: it should be accessible to a reader not intimately familiar with the topic treated;
   - a subject index: as a rule this is genuinely helpful for the reader.
   - For evaluation purposes, manuscripts should be submitted as pdf files.

6. Careful preparation of the manuscripts will help keep production time short besides ensuring satisfactory appearance of the finished book in print and online. After acceptance of the manuscript authors will be asked to prepare the final LaTeX source files (see LaTeX templates online: https://www.springer.com/gb/authors-editors/book-authors-editors/manuscriptpreparation/5636) plus the corresponding pdf- or zipped ps-file. The LaTeX source files are essential for producing the full-text online version of the book, see http://link.springer.com/bookseries/304 for the existing online volumes of LNM). The technical production of a Lecture Notes volume takes approximately 12 weeks. Additional instructions, if necessary, are available on request from lnm@springer.com.

7. Authors receive a total of 30 free copies of their volume and free access to their book on SpringerLink, but no royalties. They are entitled to a discount of 33.3 % on the price of Springer books purchased for their personal use, if ordering directly from Springer.

8. Commitment to publish is made by a *Publishing Agreement*; contributing authors of multiauthor books are requested to sign a *Consent to Publish form*. Springer-Verlag registers the copyright for each volume. Authors are free to reuse material contained in their LNM volumes in later publications: a brief written (or e-mail) request for formal permission is sufficient.

Addresses:
Professor Jean-Michel Morel, CMLA, École Normale Supérieure de Cachan, France
E-mail: moreljeanmichel@gmail.com

Professor Bernard Teissier, Equipe Géométrie et Dynamique,
Institut de Mathématiques de Jussieu – Paris Rive Gauche, Paris, France
E-mail: bernard.teissier@imj-prg.fr

Springer: Ute McCrory, Mathematics, Heidelberg, Germany,
E-mail: lnm@springer.com

Printed in the United States
By Bookmasters